SCIENTIFIC INQUIRY IN PHILOSOPHICAL PERSPECTIVE

CPS PUBLICATIONS IN PHILOSOPHY OF SCIENCE

Edited on behalf of the
Center for Philosophy of Science
University of Pittsburgh

by

SCIENTIFIC INQUIRY IN PHILOSOPHICAL PERSPECTIVE

Edited by

Nicholas Rescher

UNIVERSITY PRESS OF AMERICA

LANHAM • NEW YORK • LONDON

4720 Boston Way
Lanham, MD 20706

3 Henrietta Street
London WC2E 8LU England

Printed in the United States of America

British Cataloging in Publication Information Available

Library of Congress Cataloging in Publication Data

Scientific inquiry in philosophical perspective.

(CPS publications in philosophy of science)
Bibliography: p.
Includes index.
1. Science—Philosophy. I. Rescher, Nicholas.
II. Series.
Q175.S4235 1987 501 86-26751
ISBN 0-8191-5798-8 (alk. paper)
ISBN 0-8191-5799-6 (pbk. : alk. paper)

CPS PUBLICATIONS IN PHILOSOPHY OF SCIENCE

Co-published by arrangement with the
Center for Philosophy of Science
University of Pittsburgh

All University Press of America books are produced on acid-free
paper which exceeds the minimum standards set by the National
Historical Publication and Records Commission.

PREFACE

For the most part, these essays originated in a series of events scheduled during the 1985–86 academic year to celebrate the 25th anniversary of the founding of the Center for Philosophy of Science at the University of Pittsburgh. These activities, which were supported by the Richard King Mellon Foundation, involved many colleagues from other universities who had been involved with activities of our Center for a long time. It was a happy occasion for us to share our interests with able scholars who are also old friends.

Nicholas Rescher
Pittsburgh, PA
June, 1985

CONTENTS

IS SCIENTIFIC DISCOVERY A TOPIC IN THE PHILOSOPHY OF SCIENCE?

HERBERT A. SIMON

SOME twenty years ago, when the University of Pittsburgh Center for the Philosophy of Science was only five years old, I spoke here on essentially this same topic, although under a different title.[1] I am pleased to be given this second chance, especially since the twenty years that have flowed by have provided me with a much sharpened thesis and with a mass of new empirical evidence to support it.

The question posed by the title of my present talk arises from the distinction made by Reichenbach and others between two "contexts" or aspects of theory construction: the context of discovery and the context of justification. Philosophy has devoted much effort to studying the context of justification: "how," in Reichenbach's terms, "one rationally supports whatever he has come to think of." But the other context, the context of discovery, has been mainly neglected. Indeed, it has generally been questioned whether philosophy is properly concerned with "how one comes to think of something." Is this a philosophical question, or only a psychological question? Is there a logic of scientific discovery?[2]

Certainly there is no such thing as a *deductive* logic of discovery. Even finding proofs for theorems, in logic or mathematics, is an inductive (or abductive, if we follow Peirce), not a deductive, process. The theorem, once found, is a deductive object, its validity to be tested by its conformity to the laws of deduction. But finding it is a matter of more or less heuristic search, usually without guarantees that the search will be successful.

Nor have philosophers been successful in arriving at an agreed-upon inductive logic for the process of discovery. Inductive logic, in its many versions, is looked upon as a tool for verifying or testing discoveries, or of assigning to them a probability or level of warranted confidence. It is not viewed (at least since Bacon and Mill) as a tool for generating hypotheses.

If a logic of scientific discovery is not a deductive logic nor an inductive logic, then what is it? I should like to argue that it is a principled method, more exactly, a collection of principled methods, for generating hypotheses.[3] To ask whether such principled methods exist is equivalent to asking whether scientists have valid reasons for behaving as they do.

Einstein thought not. Einstein believed that creative discovery was accomplished "by intuition based upon something like an intellectual love of the objects of experience." As a poetic expression of the feelings that all of us experience in the face of creativity (our own or others'), Einstein's description is unexceptionable, even moving. But as an explanation, it is, alas, mystical and magical—unless and until we provide operational definitions of "intuition" and "intellectual love" as reasonable processes. In the course of this paper, I shall attempt something like that—at least a scientific characterization of intuition.

I. Scientists have reasons for what they do

In the second volume of the Reno Conference on scientific discovery (p. 48), Noretta Koertge makes a useful distinction:

> We should be very careful to distinguish among inference patterns which are legitimate in all possible worlds (these I would call part of logic), those which seem to be heuristically successful in our world (these I would call rational) and those which are typical of human thought but have no obvious systematic benefit for science.

I shall be arguing for Professor Koertge's second category of inference patterns: those that represent the reasons scientists have for what they do. What scientists do is reasonable, not in the sense that it is optimal behavior but in the sense that it is often successful in reaching its goal of discovering new theories. It satisfices. Moreover, the principles that underlie this reasonableness may not be valid in all worlds, but they are valid in a world that is lawful (patterned) and, in some sense, "simple."

If scientists have reasons for what they do, then it becomes an important philosophical problem in the context of discovery to formulate an effective heuristic of research—to guide and advise scientists in their reasoning. Such a heuristic constitutes a logic of discovery.

The valid reasoning of scientists (the logic of discovery) is a special case of human reasoning in general. Today, we have learned to characterize human reasoning as a process of heuristic search. In the case of discovery, the search is carried out by *generating* and *testing* hypotheses. But we should not suppose that the phases of generating and testing are wholly separated, the one being performed before the other. We shall see, on the contrary, that generating and testing are continually interwoven, the exploration of the space of possibilities being continually guided by evaluation of what has been found thus far. In this picture of the reasoning process, the contexts of discovery and verification merge.

Thus, I shall hypothesize that scientific discovery is a form of problem solving; and without pausing to indicate how this particular hypothesis was discovered, I should like to report some of the evidence that we have today for its validity. My remarks will be about the *process* of discovery, but they will lie in the *context* of verification of a theory of that process.

II. Where do good hypotheses come from?

Sometimes discovery is discussed as if there already existed (in Plato's heaven of ideas?) a pool of candidate hypotheses, and that hypotheses were somehow drawn from this pool to be tested. That view, of course, simply raises the question of the origins of the pool. And if we admit that such a pool exists, then the "real" discovery took place at its creation, for all that scientists can do subsequent to that creation is to test hypotheses that are drawn from it. There appears to be no room for genuine discovery.

This difficulty is best overcome by avoiding it at the outset—by denying that a pool of hypotheses is given to science. Of course we cannot manufacture hypotheses out of whole cloth; there must be some starting point. But the starting point need not be a set of full-blown theories. Instead, it may be (and, I would argue, is) a relatively small set of primitive concepts that can be enlarged by recursive combinatorial processes to generate others. Just as the vast number of different protein molecules all are made up of sequences of the twenty amino acids, with no upper bound on their variety, so scientific theories can be made manufactured by combinatorial processes from a few primitives. I will illustrate presently just how this can be done.

But combinatorial processes can produce an embarrassment of riches; they can lead to an exponential explosion of hypotheses. If that explosion is to be avoided, the generative process must be guided. The information that provides the guidance, and leads the generation in relevant directions, can come from the data that we are seeking to explain with our new theories (data-driven discovery), or it can come from previously existing theories (theory-driven discovery), or it can come from some combination of these. Let me begin by discussing data-driven discovery, and then turn to theory-driven discovery.

III. Data-driven discovery

I shall base the initial discussion of data-driven discovery on the computer program, BACON, that was developed in collaboration with my colleagues Pat Langley, Gary Bradshaw, and Jan Zytkow.[4] This is a program that, when provided with data, is capable of discovering regularity and pattern that is hidden in the data. The way in which it generates hypotheses, guided by the data that are given to it, can be illustrated by Kepler's Third Law of planetary motion.

At the time of Kepler, the periods of revolution of the planets about the Sun were known quite accurately, and their distances from the Sun with less accuracy. These two parameters summed up their aggregate behavior abstracted from the details of their orbits. For a long time, Kepler had known that the periods increased with the distances, and indeed, had conjectured that they might vary with the squares of the distances. This hypothesis fit the data rather indifferently. After apparently putting aside the question of the relation for a decade, Kepler, returned to it, and after a few weeks of work, concluded that the periods increased with the three-halves power of the distances, and found that he now had a very good fit to the data.

We do not know how Kepler generated either of his hypotheses, but we can show at least one path to them, a path that is followed by the program BACON. BACON at first has no theory about the phenomena under examination to guide it in its generation of hypotheses. However, it does have some general, content-free heuristics which direct the generation in plausible directions so as to postpone, if not wholly avert, the exponential explosion of search.

BACON constructs functions of the two observed variables, then looks to see if these functions are constant. If any of them is, it has found an invariant—a scientific law connecting the variables. If they are not constant, it adds each new function to the set and looks in the same way for relations between the new variables (theoretical terms) and the orginal ones (observables). It halts when if finds an invariant, or when it has searched for a specified time without success. The functions that BACON tests are not generated randomly, but in response to a simple heuristic: if two variables co-vary, introduce their ratio as a new variable; if they vary inversely, introduce their product as a new variable and test it for constancy. This generator of hypotheses can be generalized slightly by examining linear relations between the variables instead of their ratios. It can, of course, be generalized in other ways, but the simple computation of ratios suffices to discover Kepler's Third Law.

Thus, in this simple version of BACON, just two heuristics are employed: generation of a new function relating co-varying variables, and recursion on the new theoretical terms that are thereby introduced. As applied to the data on the periods (P) and distances (D) of the planets, BACON first notices that P increases with the increase in D. It then computes P/D, which is found not to be invariant, but which is retained. Now P/D is found to co-vary with D, and their ratio is computed, giving P/D^2, which is also not a constant. Notice that BACON's second conjecture is the same as

Kepler's initial, and erroneous, hypothesis that the periods vary with the squares of the distances. Since BACON contains a parameter that specifies the amount of variation in the values of a variable it will tolerate in its test for approximate constancy it would accept this hypothesis if the parameter were set sufficiently loose, but reject it if it were set tighter—even, we might say, as Kepler did.

Having rejected the hypothesis that P/D^2 is a constant, BACON notices that this new theoretical term varies inversely with P/D. Therefore, it multiplies these two theoretical terms, obtaining P^2D^3, which indeed, equals a constant, K_1. Thus, after examining only three hypotheses, BACON has discovered the three-halves power law, which is seen to fit the data well.

Now exactly the same procedure can be followed with data on the periods and distances of the satellites of Jupiter in relation to that planet, in which case the same law will be found, with a new constant, K_2. In fact, the procedure can be abbreviated, for BACON will use the law it has already found for the Sun to conjecture the same hypothesis for Jupiter. It is now driven by previously discovered theories as well as by data. Notice that the constants, K_1 and K_2 represent a new theoretical concept, a concept that we usually call *gravitational mass.*

Thus BACON, given only information about observables, generates not only laws but also new theoretical concepts. These concepts are of two kinds: variables like P/D that are definable in terms of observables, and variables like K that are computable from the observations, but that are not definable in the sense of Tarski, and that are associated as properties with nominal variables like Sun and Jupiter. We have here a constructive answer to the doubts expressed by Carl Hempel[5] and others as to whether new concepts can be hypothesized by a mechanical discovery system. In experiments with other empirical data, BACON has also discovered and introduced the concepts of inertial mass, specific heat, index of diffraction, molecular weight, and atomic weight.

Thus, in BACON we have a program that, using a few simple heuristics, discovers scientific laws and new concepts by means of a highly selective search. In the course of that search it generates an indefinite number of new hypotheses by applying combinatorial operators to those it already has, taking as its primitives the observables.

We turn now to a second example of data-driven discovery where our data come not from a computer simulation but from the historical record (and an informal experiment). The discovery by Planck, in 1900, of his law of blackbody radiation was a milestone in the history of physics, the first long stride toward quantum mechanics. Few discoveries in science have had consequences so momentous, and so little anticipated by the discoverer.

The phenomena of blackbody radiation had long been of interest to physicists, and the search for a law relating the intensities of the various frequencies of radiation to the temperature had been begun by Helmholtz in the 1860's. By 1896 a law , Wien's Law, had been found that fit the available data very closely. It was essentially a simple exponential, and it was discovered in data-driven fashion by fitting functions to the observations. The young Planck undertook the task of constructing a physical model of a black body that would yield Wien's law as a consequence. He demonstrated to his satisfaction that a black body viewed as a system of resonators would necessarily produce radiation in conformity with Wien's law; and he published his demonstration early in 1900.

Alas, the progress of empirical research was unkind (temporarily) to Planck. Improvements in instruments for measuring radiation in the infrared range extended the spectrum for blackbody radiation, and the data in the new range of observations departed conspicuously from the predictions of Wien's Law. On October 7 of that same year, 1900, a colleague and his wife came one Sunday afternoon to call on the Plancks. In the course of their conversation, the colleague reported data he had recently obtained that showed conclusively that Wien's Law did not fit radiation in the infrared range.

Planck was now in an embarrassing position. His proof of the "necessity" of Wien's Law was confronted with data that demonstrated its "impossibility." His first response was to search for a new law that would fit the data over the entire range of observations. He completed that search on the same day of October 7 and posted a card to his friend reporting the new function—the one we now call Planck's Law of blackbody radiation.

But there still remained the task of finding the flaw in the "proof" of Wien's Law, and correcting it. That task took Planck several months of what he later described as "the hardest work in my life." This work was driven both by data—the empirically valid law—and by theory—classical mechanics and thermodynamics, as well as Boltzmann's very recent developments in statistical mechanics based upon probability theory. In the course of the revision, Planck made an unorthodox (i.e., un-Boltzmannian) assumption about the underlying probability distribution—an assumption that was forced upon him by the result he had to obtain. Out of that assumption came the quantum.

I should like to attend, however, to the first, data-driven, portion of Planck's work: his discovery on October 7 of the empirical law. Clearly, this part of his work was driven wholly by the data, without theoretical motivation. In the spirit of casual empiricism, I conducted an experiment in the lunchroom of Carnegie-Mellon University. Over a period of days, I sat down at lunch with a number of different colleagues whom I knew to be good applied mathematicians. I said to them (in the deceptive way of psychologists): "I have some beautiful smooth data, but I can't find a function to fit them. For large values of x, y is expontential in x; but for small values of x, y is linear in x through the origin. Can you suggest a function that would fit the data over the whole range?

Five of my eight colleagues, in less than two minutes each, produced a function equivalent to Planck's Law. (Essentially, it involves replacing the e^x in Wien's Law by $e^x - 1$.) Some of them, after they had obtained the result, noticed its resemblance to Plank's Law, but none were reminded of Planck's Law when I stated the problem to them. Now my colleagues, alas, did this in 1984 instead of 1900, so they will not be rewarded with a Nobel Prize. But their response provides strong evidence that a simple data-driven discovery system, knowledgeable about standard mathematical procedures (in this case, the simplest way to the result involves expanding e^x into a Taylor's series), would reach the same result that Planck did. The evidence from Planck's October paper on his discovery indicates that he also arrived at it along a data-driven route, but a slightly different one from the Taylor's series expansion that some of my colleagues employed.

The conclusion I would draw from the experiment is that there exist so many plausible discovery heuristics for reaching the goal that many of those who are "skilled in the art"—i.e., are in possession of a subset of these heuristics—will find, quite expeditiously, the law that fits the data. All of them need not use identical heuristics, for a number of different paths will lead them to the same goal.

I cannot offer a comparably detailed picture of how Planck, in the months from October to December, 1900, arrived at a physical explanation for the empirical law. From what has already been said, it is clear that he worked backward from the known conclusion and his previous (erroneous) explanation, probing for the weakest points in the explanation and modifying it in a way that would lead to the required result. The unorthodox probability assumption (and I do not offer an explanation of how that was generated) provided the necessary link in the new derivation. Whatever the details of the process, no one would describe it as random or unreasoned.

IV. "BUT THE REAL DISCOVERY IS . . ."

Not everyone accepts BACON's derivation of Kepler's Third Law as an adequate explanation of the discovery. "Of course," they say, "If BACON is told specifically to relate the periods of the planets with their distances from the Sun, it will find the law. But the *real* discovery of Kepler consisted in selecting these variables as the relevant ones." As we shall see, there are other candidates besides the screening and choice of variables, for the *real* discovery, but this is one that is often proposed.

There are two difficulties with the objection: one specific and one general. The specific difficulty is that in the case of Kepler's law, there did not exist any very large set of variables from which to select the relevant ones. If we take the orbits as approximately circular, then the period and the distance constitute what we would today call "sufficient statistics" for describing the planetary motions. They are the obvious parameters for characterizing those motions. Probably, the third parameter we would select would be the eccentricities of the orbits. And even if these and other parameters were included in the set provided to BACON, BACON's heuristics would quickly find the others to be irrelevant and would simply ignore them. It would not be greatly slowed down in its discovery by their presence.

The general difficulty with the objection that BACON has not made the *real* discovery, is that no single step in the discovery process deserves that label. The moment of discovery is simply the final one of a whole sequence of activities and events, each of which is to be explained by the conditions that immediately preceded it. In this respect, a theory of scientific discovery is no different from any other dynamic theory that explains events in terms of differential or difference equations. The explanation consists in showing that if the current situation, at any given moment, were such and so, the next step would follow according to the equations. The "equations," in BACON's case, constitute a computer program—quite literally a set of difference equations, whose time interval is the instruction cycle time.

If no single step in the discovery process stands as the *real* discovery, then there are many questions that need to be answered to account for any single incident of discovery. BACON answers mainly the question of how a discovery can be induced once the data are given. We can also ask (although in this particular case of Kepler's Law it is not a very interesting question) how these particular variables were chosen for consideration. We can ask where the instruments came from that were used to make the observations. We can ask what role intuition and happy accidents played in the discovery process. And we can ask how the problem came to be represented as it was.

A comprehensive theory of discovery has to provide answers to all of these questions, and perhaps to others. Today, we cannot answer them all in the detail that is offered by a running computer program, but we can give at least general accounts of the processes that go on. We need not regard these components of the discovery process as more mysterious or magical than the induction of laws from data.

The Choice of Data. The choice of variables to observe is often no more problematic than it was for Kepler. A scientist working in any given domain finds that a few variables have already been established as the important and measurable ones, and he or she generally works with these. When Joseph Black undertook to understand temperature equilibrium in liquids and solids, he had balances for measuring weights, containers for measuring volumes, and thermometers for measuring temperatures. Even the experiments he performed were not new—they had been done by Fahrenheit and

Boerhaave thirty years before. Choice of data was simply not a question he had to face.

The Invention of Instruments. But where did Black's instruments come from? How was the thermometer invented? The answer, in this case and more generally, is that instruments usually evolve from the discovery of phenomena. After it was discovered that many substances expand when they are heated, the expansion phenomenon was translated into a method for measuring degrees of heating. Tswett discovered that mixtures of closely related organic substances stratified when filtered through a column of clay particles. He turned this discovery into the method of chromatography. Ohm discovered that the current in a circuit was inversely proportional to the length of a wire inserted in it, thereby giving rise to an instrument for measuring resistance. The list is very long, and a great many instruments can be accounted for in this way.

The Origin of New Phenomena. But perhaps the discovery of new phenomena is the *real* creative act. Where do new phenomena come from? One source is the systematic design of experiments that extend observations of well-established variables into new ranges. We have already mentioned one example: the extrapolation of blackbody radiation into the infra-red range (which, in turn, depended on the invention of improved bolometers).

In fact, a major application of new instruments is to use them to search for new phenomena that could not previously have been observed: elementary particles with accelerators of higher energy, distant celestial objects with more powerful telescopes, atomic structures with chrystal diffraction techniques.

But an enormous fascination attaches to those discoveries where "accident" appears to play an essential role—where a phenomenon is observed that was not anticipated and was not even being searched for. Fleming noticed that his bacteria were dying on a Petrie dish on the edge of which was a growth of mold. Fleming had, in the words of Pasteur, a "prepared mind." He knew how bacteria should look, and how they looked when destroyed by lysis. He knew a mold when he saw one, and he knew that mold didn't belong on his Petrie dish.

Thus are produced the conditions of surprise: a set of well defined expectations of a prepared mind that are disappointed by the reality. Surprise is followed by attention to the phenomenon: lysis of bacteria by mold. If the event is sufficiently interesting (in terms of the knowledge and values of the scientist), attention is followed by the design of experiments that will exploit the surprising phenomenon systematically. What range of bacteria can be lysed? What range of molds are capable of lysis? What is the speed of the process with each pair of ingredients? Can the effective agent be isolated: If the mold is squeezed or crushed or distilled, and the product purified chemically, does it still have the power of lysis? These are the standard experimental operations that led to the discovery of penicillin by Fleming and Florey.

The "reasonableness" of the steps in a program of experimentation becomes obvious when we have available a detailed laboratory log of the experimenter's work. Such a log for the laboratory of Hans Krebs, during the period of his work on the urea cycle, has been analysed by Frederic L. Holmes.[6] The program of research began with the desire to exploit a new technique—the study of metabolic processes in tissue slices—which Krebs had learned as a postdoctoral student in the laboratory of Otto Warburg. He selected the problem of urea synthesis because it was a problem of importance on which a good deal of work had already been done with other techniques.

The question to be answered was where the nitrogen in urea came from—from ammonia or from amino acids—and by what chemical route. This question did not have to be invented by Krebs; it had been asked long before. The method of searching for an answer was to test various combinations of amino acids and ammonia in liver tissue slices, and measure the amount of urea produced. The surprise came when combinations of a particular amino acid, ornithine, in combination with ammonia, produced large quantities of urea. Further experiments showed that the role of the ornithine was catalytic, and inferred a plausable reaction path.

Problem Representations. Finally, we may ask where problem representations come from. Does the *real* discovery inhere in devising a clever new representation that makes patent the inner structure of the phenomena: Faraday's lines of force, or Bohr's planetary model of the atom? Expressing a problem in terms of an appropriate representation is certainly an important part of the scientific discovery process, but is no more unique as a step than any other kind of step that we have discussed.

In the case of Faraday, the representation came from the phenomena themselves—the visible arrangement of iron filings in the presence of a magnet. But more generally, it may be said that scientific discovery seldom, if ever, calls for the invention of a totally new representation. The repertory of representations for phenomena and theories is relatively limited, and the scientist's task, if a representation is needed, is to select one from among those already available. Thus, systems that are distributed in space can be represented as systems of particles, as wave fields, as combinations of these, and in few, if any, other ways.

The basic representations, in fact, have very long histories, most of them extending back at least to classical Greece. Certainly particles and waves have been with us since then, were inducible from common, everyday phenomena, and have played a central role ever since in the representation of phenomena in all of the sciences.

There are obviously some important questions that remain to be answered about where representations come from. A good starting point would be to study the processes whereby scientists select particle or wave models to represent new phenomena—for example, Mendel's selection of particles, in the form of genetic "factors," to represent the elements of his theory of inheritance. But the process has not yet been studied extensively, and I cannot elaborate on it further in this discussion. I would only argue that a theory of scientific discovery does not need to include processes for making representations out of whole cloth. No such processes seem to be discernible in the actual history of scientific thought.

V. Conclusion

What can we conclude from our brief survey of the variety of activities that are involved in the development of new scientific knowledge? I think we can conclude that the processes of discovery are essentially the same processes that we observe in other forms of problem solving. We can conclude, further, that there are better and worse ways of problem solving; hence that it is possible to construct normative theories of problem solving and of scientific discovery.

I have tried to illustrate a small fragment of such a theory as it is embodied in the heuristics of BACON, and at a less formal level, in my comments on where problems come from, how instruments are invented, how new phenomena appear, and how representations are formed. These heuristics take the form of if-then rules that allow scientists, on recognizing familiar features in situations with which they are confronted, to choose reasonable steps to advance theory or factual knowledge. Among the heuristics that have been proposed here (always by way of example) are:

- If you have data, apply the BACON heuristics. (You may find a law.)
- If you are surprised, find the scope and limits of the surprising phenomenon, and examine the causal mechanisms that underlie it.
- If you discover a phenomenon, transform it into a new instrument of measurement.
- If you have a theory, apply it to as many phenomena as possible (and look for surprises).

All of these rules are heuristic. On many occasions, they may not work. They will not discover all laws. (But do we have any reason to believe that is possible?) They will not discover laws in data that are not lawful.

But the heuristics will perform a lot better than random search; they are much superior to depending on the laws of chance for discovery. They are the core of a normative theory of scientific discovery that makes no guarantees, but that, by way of compensation, requires no miracles and postulates no mysteries.

Such heuristics lie at the core, too, of a theory of human reason that gives up illusory concerns with certainty and optimization. Instead, the theory describes human reasoning as an heuristic satisficing process that enables human beings to engage, sometimes, in successful thinking.

NOTES

1. The talk is reprinted in two parts in my *Models of Discovery* (Dordrecht: Reidel, 1977), pp. 268–303, under the titles, "Thinking by Computers" and "Scientific Discovery and the Psychology of Problem Solving."

2. This topic is also discussed in *ibid.*, pp. 326–337.

3. Perhaps this is what Peirce and Hanson *should* have meant by "abduction." Whether this actually was their meaning will not be examined here.

4. See Langley, Simon, Bradshaw, and Zytkow, *Scientific Discovery* (Cambridge, Mass.: MIT Press, 1986).

5. See his "Thoughts on the Limitations of Discovery by Computers," in K.F. Schaffner (ed.), *Logic of Discovery and Diagnosis in Medicine* (Berkeley: University of California Press, 1985), especially pp. 115-116.

6. This account is drawn from his "Hans Krebs and the Discovery of the Ornithine Cycle," *Federation Proceedings, Federation of American Societies for Experimental Biology*, vol. 39 (1980), pp. 216–225.

TOWARD A COGNITIVE THEORY OF SCIENCE

RONALD N. GIERE

I. Why do we need a cognitive theory of science?

THE year 1960 was quite special for the philosophy of science. The Center for Philosophy of Science at the University of Pittsburgh, the Boston Center for Philosophy of Science, and the Department of History and Philosophy of Science at Indiana University were all founded that year. All three institutions were to some extent inspired by and modeled on the Center for Philosophy of Science founded several years earlier by Herbert Feigl at the University of Minnesota.[1]

In retrospect, 1960 can also be seen as one of the watershed years for the philosophy of science as a whole. The previous half century had been dominated by the founding, development, and elaboration of Logical Empiricism. The modern return to a more historical approach to the philosophy of science was just beginning. I suspect that many who took part in this "historical turn" could not help thinking of themselves as the vanguard of a Kuhnian revolution in the philosophy of science, or, somewhat later, as part of a progressive problem shift. As it turned out, there was no Kuhnian revolution. Rather, we now have two research traditions—one "logical" or "analytical," the other "historical"—whose interactions range from rivalry to indifference.

After much reflection, I have concluded that this bifurcation has not produced a healthy dialectical tension, but an unhealthy schizophrenia for which a cure is needed. Moreover, it seems to me unlikely that a cure will emerge internally as the result of some dialectical process. It must be found somewhere else. But where?

My suggestion is that we look to the newly emerging cognitive sciences. Science, after all, is itself a preeminent cognitive activity. And the cognitive sciences are developing a set of concepts and methods—and results!—which may provide a solid basis for the study of science regarded as a natural, human, cognitive, enterprise.

Here one must be careful not to take too narrow a view of what constitutes cognitive science. Some would equate it with cognitive psychology. Others would restrict it to the application of computational models, whether to humans or to computers. I would insist on a broad view that runs all the way from cognitive neurobiology, through cognitive psychology and artifical intelligence, to linguistics, and on to cognitive sociology and anthropology. One must not put a priori restrictions on what might be useful in understanding the phenomenon of modern science.

On this suggestion, the role of these new concepts and methods in a cognitive study of science would be analogous to the role of logical concepts and methods in the logical study of science or to the role of historical concepts and methods in the historical study of science. But a thoroughgoing cognitive study of science would absorb, modify, or replace these latter concepts and methods. In particular, the history of science becomes the cognitive study of *past* science. And the theories of analytic philosophers of science become themselves scientific models to be tested against the phenomena of scientific practice just as the models of, say, economic theory, are to be tested against the phenomena of economic performance.

II. Justification and Rationality

The idea of using the concepts and methods of the cognitive sciences to understand science itself is not new. To cite just one example, the work of Herbert Simon (1977, 1986) and his associates on scientific problem solving and scientific discovery is well known and respected among philosophers of science. But such work has not been regarded as a prototype for a more general philosophical approach to the study of science. The reason, I think, is that this work has been seen has having a different aim than that of the philosophy of science.

Philosophers have generally assumed the goal of showing that the claims of science are *rationally justified*. Or, to put the same point somewhat differently, the philosophy of science has tried to do more than merely *describe* the actions of scientists. Its goal, it has been claimed, is to provide an account of *good* science and thus to *prescribe* the actions scientists should take.

Outside of philosophical circles, the immediate answer to questions about the justifiability of the claims of science is that they are warranted by the obvious success of modern science. Thus, if one wants to give advice on how to do good science, one must first investigate empirically what makes the successful sciences so successful, and then use this knowledge as the basis for advising others how to do likewise. There is nothing more to being "rational" than simply using one's knowledge effectively to achieve one's goals.

The reason this "obvious" answer has not generally been accepted by philosophers is that it is *circular*. It appeals to the success of science. But the original question is interpreted as asking whether the claims of science to be successful are in fact justified. To understand this charge of circularity it is helpful to contemplate briefly the prototype of circularity in modern philosophy, The Cartesian Circle.

Descartes argued for the existence of god on the basis of having a clear and distinct idea of a perfect being that exists necessarily. He then went on to justify the method of clear and distinct ideas by arguing that a perfect being would not allow one to be deceived by a clear and distinct idea. But this, his critics objected, is circular. The objective is to give a "straight" argument from some indubitable premises to the desired conclusion. The assumption, of course, is that such a "straight" argument is possible.

Before the scientific revolution it was among the main goals of philosophy to justify theology. Many later philosophers concluded that science is better than theology and took it upon themselves to justify science. The object to be legitimated changed, but the goal and the methods remained the same: that of providing a non-circular argument from secure premises. What is surprising is that so many people who would reject the task of providing this sort of justification for theology nevertheless thought that the same goal might be achieved for science.

The enterprise of trying to justify science without appeal to any even minimally scientific premises has been going on without conspicuous success for three hundred years. One begins to suspect the lack of success is due to the impossibility of the task. Perhaps there just is no place totally outside science from which to justify science. At the very least one might conclude that the task is not going to be accomplished any time soon by ordinary mortals. Of course the skeptic would question the justifiability of this claim as well. But I am willing to take it as sufficient grounds for an ordinary mortal to try something else.

III. Evolutionary naturalism

I have suggested that the cognitive sciences may provide appropriate concepts and methods for studying science as a human activity. Yet there is a still more general empirical warrant for this approach which is provided by evolutionary theory. Human cognitive capacities evolved through natural selection. This includes, in particular, our capacities for perceiving and representing the world, for coordinating our movements in it, and for communicating with our fellow humans. There is thus a solid basis for understanding how we come by *some* of our knowledge of the world, namely that which was required for survival and procreation. The details of this process are now the subject of intense scientific investigation. Just the barest outline, however, is enough to throw considerable empirical light on traditional philosophical debates between empiricism and rationalism.[2]

The evolutionary picture shows that the starting point for empiricism is just plain mistaken. Individuals do not build up their world from sense impressions. Rather, through natural selection our perceptual mechanisms have become adapted to process the kind of information the world in fact presents to us. The operation of these mechanisms is largely opaque to our conscious attention. Our eyes, for example, already process the changing perspectives of a moving animal as the continuous motion of a single object. We do not infer or construct it from a series of visual sense data.[3]

Kant, therefore, was right that our perception of objects as moving in roughly a three dimensional Euclidian space is built into the very way we perceive objects. But he was wrong in seeking a transcendental basis for this fact. It is just the adaptive response of an organism evolving its perceptual machinery in a Euclidian environment where its survival and reproduction depend on being able to interact effectively with medium sized objects, some of which, like themselves, are self-mobile. Moreover, there is no empirical reason why such an organism should not develop the representational capacity to conceive of a *non*-Euclidian space.

Of course the traditional skeptic would wish to raise the charge of circularity and challenge the justifiability of our claims about evolutionary biology. But unless such skeptics can show how a straight justification is possible, we need not take them seriously, any more so than we do children who respond to every explanation with a further, "And why is that, Mommy?" The evolutionary picture provides a framework in which we can explain how we acquire our scientific knowledge, including our knowledge of evolutionary biology. The interesting task is to fill in the details of this process.

Some recent advocates of evolutionary epistemology, such as Campbell (1974), would go further than just appealing to the evolution of our general cognitive capacities. Some suggest an evolutionary basis for more specific dispositions, such as a tendency to infer the existence of a common cause behind repeated coincidences (Ruse, 1986). Others would apply evolutionary models to knowledge itself, as in Toulmin's (1972) classic attempt to construct a theory of evolving populations of concepts. Whether such claims might turn out to be correct is entirely an empirical matter. I, however, will not appeal to any such doctrines. For me, evolutionary biology mainly plays the kind of background role that empiricism played for philosophy of science in the days of Logical Empiricism. And just as Logical Empiricists turned to logic for their more specific concepts and methods, I turn to the cognitive sciences.

The thesis that the study of science as a human activity is best conceived as a cognitive science of science is part of a more general *Naturalism*—the view that human activities are to be understood as entirely natural phenomena just like the activities of chemicals or animals. Like any such general viewpoint, naturalism has been "refuted" by philosophers dozens of times. Putnam's (1982) attack on Quine is a recent example. What makes naturalism attractive now, however, is not that such philosophical arguments have been refuted, but the *empirical success* of the cognitive sciences. This is the historical pattern. Proponents of the new physics of the seventeenth century won out not because they explicitly refuted the arguments of the scholastics, but because the empirical success of their science rendered such arguments irrelevant.

IV. Representation and judgment

Two central issues Logical Empiricism were the nature of scientific theories and the nature of scientific inference. The standard view was that theories are interpreted formal systems set out in axiomatic form. So theories are sets of statements which may be true or false. In addition, at least some of these statements should be generalizations of universal form. Scientific inference was held to follow some form of inductive logic. Both of these issues have remained important since the breakdown of Logical Empiricism, but there is now far less unanimity on the answers.

Some philosophers of science within the recent analytical tradition now regard theories as *families of models*. Empirical claims about the world are made by means of some kind of mapping, or structural isomorphism, between a model and a real system. This view of theories has become associated with the "structuralist" program for reconstructing scientific theories in a canonical form (Stegmueller, 1979). It is also, under the heading "the semantic view of theories," associated with van Fraassen's (1980) anti-realist "constructive empiricism."

That theories are to be understood as families of models fits nicely within the cognitive sciences under the general heading of "representations." That humans (and animals) have internal representations of their environment (as well as of themselves) is probably the central notion in the cognitive sciences. It is, for example, what distinguishes cognitive psychology from behaviorism. Depending on the local environment within the cognitive sciences, one finds talk of such things as "schemata," "cognitive maps," "mental models," and "frames." Assimilating scientific models to the cognitive scientists' schemata somewhat downplays the fact that scientific models are often described using written words or symbols. It also allows for a "softer" relationship between models and the world, one more naturally described in terms of *similarity* than isomorphism.[4]

In this paper I shall be more concerned with inference than with theories as such, but I will adopt the idea that scientists use families of models. *Inference*, in the cognitive sciences comes under the general heading of "Human Judgment." Here one is not concerned with the validity or justifiability of logical rules, whether deductive or inductive, but with how humans make judgments, and with how *reliable* their judgments, might be.[5] I shall be focusing on the question of how scientists judge (or decide) whether a particular model is acceptably similar to a real world system. And I shall be particularly interested in the role of *experimentation* in such decisions.

I shall be focusing on a special area of research in contemporary nuclear physics. Although limited in scope, the results of such a study should be more widely applicable. Nuclear physics is a paradigm of twentieth century science. In its use of mathematical techniques, of computers and other advanced technology, as well as in its organization into research groups, nuclear physics resembles many other contemporary sciences, including, for example, molecular biology.

For several years I have had the opportunity to observe research at a national cyclotron facility. Thus, in addition to published work, I have had access to more "ethnographic" sorts of data—participant observation, interviews (in some cases repeated interviews with the same individual), annual reports, lecture notes, and even research proposals which contain an explicit "scientific justification" for proposed experiments. Only a small part of all this varied information will be presented here.

V. Two families of models

One of the basic questions in nuclear physics is the nature of the force experienced by a free moving proton as it interacts with a nucleus by means of an elastic collision. An elastic collision is one in which no energy, whether in the form of particles or radiation, is given off in the interaction, though of course the proton and the nucleus may exchange energy in various ways.

Up until 1980 it was generally thought that the nature of this interaction could be adequately represented by a model employing the famous Schroedinger equation. This equation plays roughly the role in quantum theory that Newton's second law plays in classical mechanics. In simple treatments, the equation contains terms representing two different forces. The strongest is an attractive force due to the average nuclear force of all the nucleons (protons and neutrons) in the nucleus. The second is a weaker force (about one tenth as strong) due to the interaction of the nucleus with the magnetic moment of the proton. This latter force arises because protons exhibit properties analogous to that of a rotating charged sphere. This property is called "spin" and the force generated is called the "spin-orbit" force.

For over a century it has been common to represent forces by a related function called a *potential*. For present purposes it is only necessary to know that negative potentials correspond to attractive forces and positive potentials to repulsive forces. The two potentials in the simple Schroedinger model are pictured in Figure 1. Here the strength of the potential is measured in million elecrton volts (MeV) and is shown as a function of distance from the center of the nucleus measured in fermies (one fermi equals 10^{-13} centimeters). Note that both potentials are attractive and that the spin-orbit potential is roughly a tenth as strong as the direct nuclear potential. The force experienced by a passing proton corresponds to the *sum* of these two potentials.

It is important to realize that the model represented by Figure 1 is just one member of a large *family* of Schroedinger models of the nuclear potential. Other members of the family have potentials with a somewhat different shape and strength. Some have additional potentials corresponding to still other forces roughly the strength of the spin-orbit force.

Around 1980 some new experiments gave results that clearly did not fit calculations derived from any simple Schroedinger model. These experiments were genuinely new in that they involved several pieces of new technology. One new machine produced a pro-

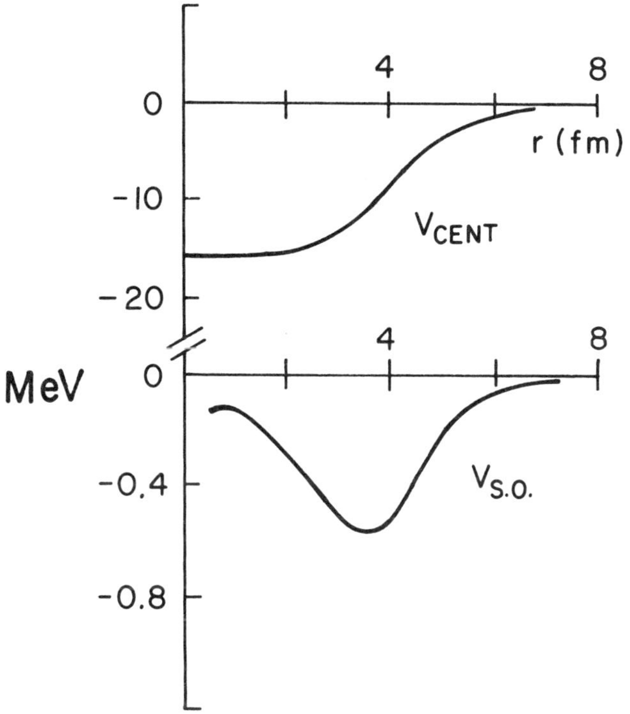

Figure 1. The central and spin-orbit potentials for a standard Schroedinger model of the nucleus.

ton beam with the spin of the protons oriented in a predetermined direction. This made possible the measurement of a parameter called "analyzing power," or *Ay*. A second new technique made it possible to measure the spin of protons both before and after their interaction with the nucleus. It thus allowed determination of the *change* in orientation of the spin of the proton as a result of its interaction with a nucleus. The resulting "spin rotation parameter," labeled *Q*, is pictured in Figure 2. One need not know anything more about these experimentally measurable parameters except that their expected values as a function of the scattering angle can be derived from any serious nuclear model.

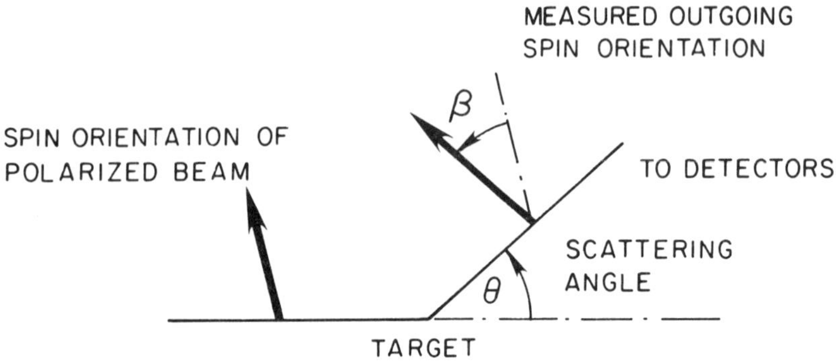

Figure 2. Schematic representation of an experiment to measure the spin rotation parameter, Q.

The immediate reaction to the new data was about what any student of the scientific enterprise would predict. It was generally thought that something must be missing in current Schroedinger models, but that other members of the Schroedinger family of models might do the trick.

If one were treat this episode as a small scale Kuhnian revolution, one would predict a period of crisis during which a variety of new models, including some not based on the Schroedinger equation, would be developed. That is not at all what happened. New models were indeed tried, but the most radical of these employed an equation that had been known for fifty years—the Dirac equation.

The Dirac equation is the relativistic version of the Schroedinger equation. Its relation to the Schroedinger equation is like that of Einstein's equations for special relativity to Newton's equations of motion. This raises an interesting question. Most physicists believe that the world is ultimately relativistic. Indeed, standard calculations of the *kinematics* of an interaction, that is, velocities, angles of scattering, and so forth, routinely employ special relativity. Yet standard Schroedinger models of nuclear *interactions* ignore fundamental relativistic effects, and in particular, the creation of new particles following the well-known relationship, $e = mc^2$. How can this be?

The answer is simple. As can be seen from the standard potential functions pictured in Figure 1, the energy involved in this sort of elastic collision is of the order of ten to twenty *MeV*. The relativistic creation of new particles, whether mesons or nucleons, requires three to five *hundred MeV*. With so great an energy discrepancy, the quantum mechanical probability of relativistic interactions is so small that it has universally been regarded as insignificant.

It is not the case that people suddenly rediscovered the Dirac equation. Rather, there were several theorists who had been developing models in the Dirac family for many years. But the number was small, "a handful of people," as one of my informants put it. This suggests the picture of competing research traditions as developed by Lakatos (1970) and Laudan (1977). In the abstract sense that Dirac models were being pursued by some people, this may be so. But I have as yet found little evidence of an *active* tradition. For example, the few people working on Dirac models seem not to have been much in contact with one another. Nor had they been actively arguing against the use of Schroedinger models. Rather, the were working on Dirac models for a variety of reasons having little to do with medium energy, proton elastic scattering. An ecological model in which a population always includes a number of variants may be more appropriate. The change in the scientific environment caused by the new data made these models suddenly relevant in a way they had not been before. Only at this point do there emerge interacting groups of people actively developing and promoting the use of Dirac models in nuclear physics.

One group scored an early success in 1982. They used the new data on analyzing power to determine numerical coefficients in a model with several Dirac potentials. The resulting model then *correctly predicted* the data on spin-rotation without any further adjustments in coefficients. By all accounts, this feat stimulated a great deal of interest in Dirac models. As one of my informants put it, "People were really shocked . . . I mean it was like, wow." Others of my informants were less enthusiastic, but all agreed that many people were impressed.

The potentials utilized in this particular member of the Dirac family of models are shown in Figure 3. As before, the total force experienced by a passing proton is the *sum* of these two potentials. But note that one of the potentials is very large positive (repulsive) while

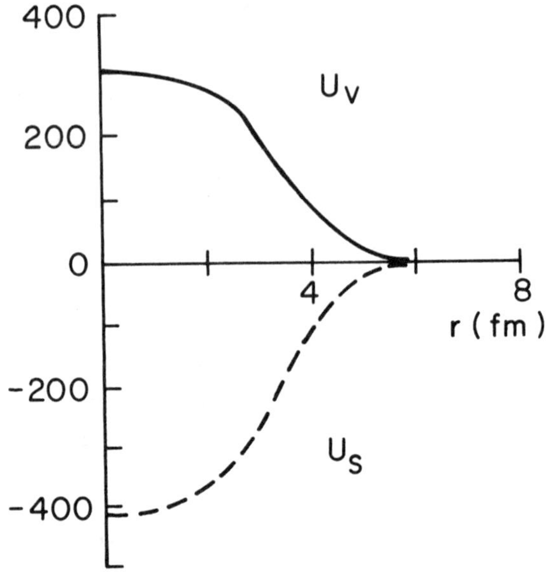

Figure 3. The vector and scalar potentials for a simple Dirac model of the nucleus.

the other is very large negative (attractive). The magnitude of each individual potential is several hundred *MeV*, and thus plenty large enough to provide a high probability for the relativistic creation of mesons or other nucleons. The *sum* of the two potentials, however, is a relatively weak negative potential like that in the Schroedinger model pictured in Figure 1.

From the standpoint of the Dirac model, earlier experiments were sensitive primarily to the net potential and not to its components. The new experiments done with a spin polarized proton beam, on the other hand, are sensitive to the individual large potentials. This is why Dirac models work better for the new data. Everyone agreed that this was a *possible* explanation of the situation. Not everyone agreed that it was *correct*, that is, that a Dirac model better represents the real interactions than any Schroedinger model.

Following the initial success of the Dirac approach there was much activity on several fronts. Some theorists devoted their efforts to developing Dirac models, others to developing better Schroedinger models. Existing data were used to judge how well these various models did in "predicting" known experimental values of *Ay* and *Q*. And experimentalists set about designing new experiments to measure *Ay* and *Q* for a greater variety of nuclei and proton energies. They also sought to make their measurements more precise.

The question I wish to pursue is how the data was used by individual scientists in reaching judgments as to which type of model best represents the nuclear potential.[6]

VI. Are scientists Bayesian agents?

Among analytic philosophers of science there is at least a weak consensus that scientists are to be understood as Bayesian agents. A Bayesian agent would assign probabilities to hypotheses and then revise these probabilities in accord with Bayes' theorem as new evidence comes in.

In our case, the Schroedinger hypothesis would be that some Schroedinger model provides the better representation of the nuclear potential. The Dirac hypothesis would be that some Dirac model provides the better representation. On a Bayesian analysis, then, one would say that initially most physicists assigned a much higher probability to the Schroedinger hypothesis, but that the new evidence began to lower the probability of the Schroedinger hypothesis relative to the Dirac hypothesis.

Bayesian agents are often described as ideally rational agents. So the claim is than scientists do and should attempt to act as Bayesian agents, although their actual performance may deviate from ideal rationality. My view is that the program of defining an ideally rational agent is just another version of the hopeless program of justifying science. The only interesting queston is whether, as a matter of fact, scientists are, at least approximately, Bayesian agenst. Only if this is so do we have a genuine Bayesian explanation of why scientists accept one model rather than another.

One thing is clear. Nuclear physicists, at least, are not *overtly* Bayesian agents. They almost never use the concept of probability, and rarely even the word "probability," when describing their epistemic attitude toward a particular model or the impact of new data on their beliefs. This is true both of published writings and of informal conversation. Of course they use the concept of probability in describing quantum mechanical processes and the expected outcomes of experiments. And on those occasions when they use data to fit the coefficients of a model, they will use a statistical measure of goodness of fit such as Chi Squared. But probability does not explicitly play the role assigned to it by a Bayesian analysis.

So if nuclear physicists are Bayesian agents, it must be at an intuitive level which is not reflected in the way they talk about their work. But there is now extensive evidence, gathered in experiments for a generation, that people in general, even those with considerable technical training (including training in probability and statistics), do not intuitively employ Bayes' theorem. I refer, in particular, to the many experiments by Daniel Kahneman and Amos Tversky.

Advocates of Bayesian methods have known about these experiments for years, but generally regarded them as irrelevant to the task of providing a *normatively* correct account of scientific inference. For a cognitive theory of science, which aims to give an *empirically* correct account of scientific activities carried on by actual scientists, nothing could be more relevant. For those who may be unfamiliar with this work, I will give just one example. This example also illustrates a point that will be important in my own account.

In the following experiment, first performed by Kahneman and Tversky in 1972 and since repeated many times by themselves and others, subjects were presented with this problem.

> A cab was involved in a hit and run accident at night. Two cab companies, the Green and the Blue, operate in the city. You are given the following information:
>
> (a) 85% of the cabs in the city are Green and 15% are Blue.
>
> (b) A witness identified the cab as Blue. The court tested the reliability of the witness under the same circumstances that existed on the night of the accident and concluded that the witness correctly identified each one of the two colors 80% of the time and failed 20% of the time.
>
> What is the probability that the cab involved in the accident was Blue rather then Green? [7]

In many experiments using variations on this question, the answer given most often is 80%, which corresponds to the reliability of the witness. The average answer is roughly the same. The "correct" answer, obtained by employing Bayes' theorem with equal initial probabilities for an accident by any particular cab, is .41. The credibility of the witness is offset by the low base rate of Blue cabs. But typical subjects ignore the base rate and make their judgement solely on the information about the witness.

There is a very interesting variation on the above experiment which substitutes the following information for (a):

> (a') Although the two companies are roughly equal in size, 85% of cab accidents involve Green cabs and 15% involve Blue cabs.

This formulation carries exactly the same base rate information as the earlier version. Yet individual answers to this problem vary greatly, and the base rate is generally no longer ignored. The average answer is around 60%, which is a compromise between the correct answer and the reliability of the witness.

The standard interpretation of this latter result is that people are more apt to make some use of a base rate if it can be fit into a *causal schema*. Here the obvious schema is the very general one that there is something about the Green cabs, or their drivers, that is causally responsible for their having more accidents.

None of this proves that scientists might not be better off if they learned to evaluate evidence as Bayesian agents. But that is a claim that must be proven empirically. It cannot be decided a priori. It is not even clear that Bayesian methods could as a matter of fact be effectively employed by human scientists in the kind of case we are considering here.

VII. Scientists as satisficers

Forty years ago Herbert Simon introduced the notion of *satisficing* to describe the decision making activities of people in organizations. He did so because he thought that standard models of economic rationality, like Bayesian models, place unrealistic de-

	SCHROEDINGER MODEL FITS BEST	DIRAC MODEL FITS BEST
CHOOSE SCHROEDINGER MODEL	CORRECT	INCORRECT
CHOOSE DIRAC MODEL	INCORRECT	CORRECT

Figure 4. Decision matrix for a nuclear physicist choosing between Schroedinger and Dirac models of the nucleus.

mands on the cognitive capacities of real decision makers. For parallel reasons, I should like to apply satisficing ideas to the choice of models by scientists.[8]

A satisficing approach requires us to understand scientists' judgments about models as instances of *decision making*—that is, deciding which model best represents the world. This seems to me to accord pretty well with the way scientists act and talk, but documenting this claim would be difficult and I shall not attempt to do so here. I shall proceed on the assumption that a decision making framework is appropriate—at least as a first approximation.

To apply this framework we must assume that our physicists have an explicit set of *options* and recognize possible *states of the world* relevant to their options. In our example, the relevant possible states of the world would seem to be: (1) that the nuclear potential is better represented by a Schroedinger model than by a Dirac model, and (2) that it is better represented by a Dirac model. The options, then, would be: (A) concluding that a Schroedinger model best represents the world, or (B) concluding that a Dirac model best represents the world. The resulting two by two decision matrix is shown in Figure 4.

One immediate and very significant consequence of thinkng about the choice of models in terms of decision theory is that the necessity for dealing with *values* or *interests* is explicit from the start. The decision theoretic framework requires that there be a value ranking attached to the outcomes of he decision matrix. So there is no question *whether* values enter into the choice of scientific models, or even *how*—they must be attached to the outcomes of the matrix in Figure 4. The question is *what* values are employed.

Assuming that we are talking about decisions by individual scientists, the values involved would be those of the scientists making decision. But clearly not all values are idiosyncratic. Many are acquired through processes of acculturaion and professionalization.

Looking at the outcomes in Figure 4, one sees that there are two possible outcomes which answer roughly to a "correct" decision. These are accepting the Schroedinger model if it fits and, alternatively, accepting the Dirac model if it fits. The other two outcomes represent "mistakes"—here one has accepted a model as fitting when in fact its rival is a better fit.

Philosophers who have conceived the acceptance of hypotheses in decision theoretic terms have introduced the notion of *epistemic* or *scientific* values. There has been, however, little agreement on just what are the epistemic values—except for *truth*. Scientists, it is said, value truth over error.[9]

I prefer to speak of "representing" or "fitting" rather than "truth," but the intent is the same. And, indeed, my formulation of the decision problem makes it difficult to imagine how any scientist could prefer either of the mistakes to either of the correct outcomes. No scientist could agree that one model was clearly a better representation of specified aspects of the world, but nevertheless prefer to accept the other as being the better representation of those same aspects. Agreeing that science is a representational activity commits one to at least this much in the way of an evaluation of outcomes in the decision problem.

It is quite clear, however, that actual scientists do not value different potentially correct decisions equally. I have interviewed scientists who claim they genuinely do not know whether or not Dirac models are really better, but nevertheless clearly prefer that correct outcome in which the Dirac model actually does provide the better representation of the nuclear potential. Such scientists must have some other values in addition to the value of being correct.

Philosophers typically seek these additional values among such scientific virtues as simplicity, precision, and wide applicability. Sociologists look to professional and other broader interests. I think *both* are correct. My informant who clearly preferred Dirac models said that if one begins with the Dirac equation the spin-orbit potential appears automatically. One gets it "for free." The Schroedinger approach provides no similar account; the spin-orbit potential is added into the equation for other reasons. This sounds like a rejection of *ad hoc* components and an appeal to simplicity and unification. Yet this same scientist had begun working on Dirac models as a graduate student—because his supervisor suggested it—shortly before they became popular, and had just completed co-authoring a text on Dirac models. It would be very difficult to show that his appeals to elegance and simplicity were not to some extent rationalizations for a commitment whose basis lay in professional training, expertise, and identity as an expert on Dirac models.

One of the major general lessons from recent research on human judgment is that humans are very poor at separating out the various factors that influence their judgments. Since scientists obviously have both professional and social interests, any model of scientific decision making which restricted consideration to some limited set of "scientific values" would stand little chance of fitting the actions of real scientists. Just as an empirically adequate theory of science cannot make unrealistic demands on the cognitive capacities of scientists, neither can it make unrealistic demands on their ability to separate epistemic from other values and interests.[10]

The problem, in short, is not to engage in a vain attempt to eliminate non-epistemic virtues from science, but to understand how science works as well as it does *despite* the inability of scientists clearly to distinguish epistemic from non-epistemic values. All this is not to say that epistemic values should be ignored. They simply must take their place along side other sorts of values or interests.

So far, then, our account of scientific decision making has the scientist valuing correct outcomes over mistakes, but otherwise free to place any value ranking or measure on the outcomes of the decision matrix. Such freedom, however, is not consistent with satisficing strategies. In order to apply a satisficing strategy, the outcomes representing correct decisions must be regarded as *satisfactory* while those representing correct decisions must be regarded as *unsatisfactory*.

This is not a trivial restriction. It is possible to imagine a scientist whose career and skills are so bound up with a particular model that to reject this model would make it impossible to go on doing science. Such a person might well regard neither outcome associated with the acceptance of the rival model as being satisfactory. The only satisfactory outcome is accepting the favored model. This might describe the situation of famous "holdouts," like Priestly, who continued to accept their old model long after the vast majority of the profession had given it up.

Let us label such scientists not as "irrational" but merely as *closed minded*. They simply will not countenance the rejection of their preferred model. By contrast, an *open minded* scientist will be one who at least regards both correct outcomes as satisfactory, even though perhaps placing a far greater relative value on one of the two correct outcomes. My research has yet to turn up a nuclear physicist who I could clearly identify as closed mined.[11]

A *satisfactory option* is one for which all the associated outcomes are regarded by the decision maker as at least satisfactory. As it stands, then, the decision matrix of Figure 4 has no satisfactory option. One outcome for each option is unsatisfactory. This should not be too surprising since we have not yet taken any explicit account of the experimental evidence. Our problem, then, is to determine what kind of evidence is required to change the decision problem so that the choice of one model might become a satisfactory option.

VIII. Satisficing and Experimental Design

What is experimentation for? A common assumption is that it is to provide *information* about the world. In the present context we can be more specific. The purpose of experiments is to provide a basis for making *decisions* about the relative fit of various models to the world. And this "basis" should reside at least partly in the world itself. It is therefore very important that experimentation involve a physical set-up which is a causal process in the world.

At least some parts of the experimental process must be open to direct, physical manipulation by the experimenters. And the process must have a known range of possible results. This range is determined by the physical nature of the set-up itself. It is a helpful abstraction to think of the parts of the process that can be manipulated as providing the *inputs* (or initial conditions) of the experiment, and of the results as being *outputs* (or final conditions).

The connection between the models under investigation and the output of the experiment is provided by the fact that a physical system of the type being modeled must be a causally relevant part of the experimental set-up. Moreover, for any model under investigation, the experimenters must be able to determine what range of outputs should result if that model in fact fits the system being modeled.

To keep things just a bit abstract for a moment, let us suppose that there are two models at issue, a "standard" model (*S*) and a "new" model (*D*). In addition, suppose that, given the specified experimental set-up, a range *RS* of possible results is a highly probable if the standard model is roughly correct. Similarly, given the same experimental set-up, suppose a range *RD* of possible results is highly probable if the new model (*D*) is roughly correct. This set-up is pictured schematically in Figure 5.

If we are to apply a satisficing strategy to the decision problem, there is one more set of conditions that must be realized by the experimental set-up. If the standard model (*S*) provides the better fit, it must be highly *improbable* that the experiment will yield results in the range *RD*. Similarly, if the new model (*D*) provides the better fit, it must be highly *improbable* that the experiment will yield results in the range *RS*.

The only remaining relevant consideration is the obvious supplemental decision rule:

> If *RS* is the actual result of the experiment, choose the standard model (*S*);
> If *RD* is the actual result, choose the new model (*D*).

Now let us reconsider the decision problem of Figure 3 in light of all these additions.

First, suppose the real world system is in fact roughly similar to

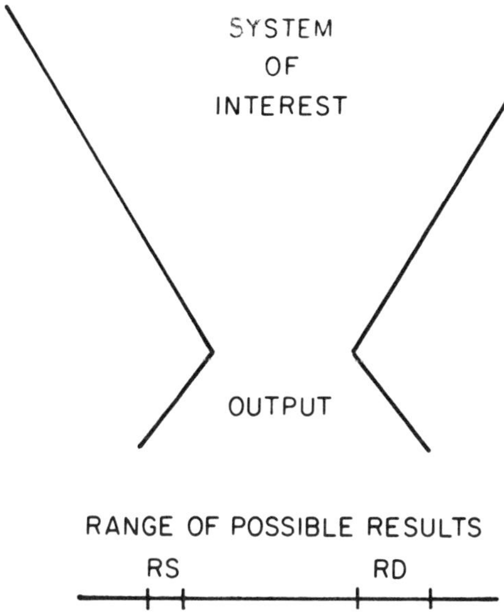

Figure 5. Schematic representation of an experiment with disjoint ranges of results for two different models of the system of interest.

the standard model (*S*). It is then highly probable that the experiment will yield result *RS*. Following the supplemental decision rule, the scientist will conclude, correctly, that the world is roughly like Model *S*. Alternatively, suppose the real world systems are in fact roughly like the new model (*D*). In this case it is highly probable that the experiment will yield result *RD*. Following the supplemental decision rule, the scientist will conclude, correctly, that the world is roughly like Model *D*.

In either case, it is highly probable that an open minded satisficer will end up with one of the two satisfactory outcomes. Correspondingly, ending up with one of the two unsatisfactory outcomes is highly improbable. Here we go beyond the original Simon model only to the extent that we regard a high probability of achieving a satisfactory outcome as itself satisfactory. This seems clearly well within the spirit of satisficing strategies.

IX. Two objections

At this point it is worth pausing to consider two objections that will likely have already occurred to the reader.

The first objection is this. Suppose that in fact neither (*S*) nor (*D*) provides a very good fit to the system in question. A much better fit is provided by a third model (*T*) which is overall much more similar to (*S*) than to (*D*). However, in the given experimental situation, (*T*) predicts results *RD*. If *RD* is the observed result, one ends up accepting model (*D*) as the better fit, when in fact (*S*) is better. Moreover, it is a matter of logic that there will always exist, it the logical sense of "exist," some such third model.

The reply is that satisficers simply do not worry about such abstract possibilities of error. If someone were actually to develop a model like (*T*), then one might try to design a different experiment to help decide between (*D*) and (*T*). Lacking any real alternative models, one accepts the possibility that one has gone seriously wrong in spite of one's best efforts.

If it turns out that scientists are indeed satisficers along the lines of the model just presented, then it would be an interesting fact about science as actually practiced that it at least sometimes succeeds in producing very well fitting models in spite of the ever present abstract possibility of serious error. This fact would require a scientific explanation. No other kind of explanation is possible. Nor should we pay heed to the skeptical conclusion that we could not possibly be justified in thinking that *any* of our models in fact provide a good fit to the world.

The second objection is this. Having just been presented with claims that humans are not in general Bayesian agents, the reader might well wonder whether the presence of probability judgments in my satisficing model does not make it subject to the same objection. I think not.

One of the results of human judgment research, exhibited, for example, in Kahneman and Tversky's experiments with the taxi cab problem, is that humans are fairly good judges of the probability of events so long as they can interpret them as the result of some causal process. That is, if they have some *causal model* which they can use in making the requisite judgments, humans are in fact pretty good judges of the probability of events—relative to the model employed.

Now the probability judgments required by a satisficing model of scientific decision making are exactly of this sort. The scientist must judge the probabilities of various possible outcomes of some definite experiment supposing that a given well-defined model is more or less correct. These are the only type of probability judgment required. In particular, the scientist is not assumed to be capable of making an inference to the probability to the model itself—the kind of inferences that typify a Bayesian approach.

X. The design and execution of an experimental test

I shall now describe some aspects of a particular experiment designed and carried out partly to test the relative adequacy of Schroedinger and Dirac models of the nuclear potential. I personally observed one run of this experiment and have interviewed some of the people involved. At several points in the experimental process one can see pretty clear evidence that the type of satisficing strategy outlined above is actually operative.

In June 1983, some months after the initial success in using Dirac models to predict the spin rotation parameter, a group of experimenters at Indiana University submitted a proposal entitled "Measurement of the Spin Rotation Parameter Q in 200 MeV Proton Elastic Scattering from C^{12} and Ca^{40}." A section titled "Model Calculations of Spin Rotation," includes a discussion of Schroedinger and Dirac models. Several members each of the Schroedinger and Dirac families of models are presented and theoretically calculated values of Q are exhibited. These are shown in Figure 6.

As one would expect, the proposal also contains considerable discussion of the experimental set-up, and of the particular equipment to be used, some of which was designed expressly for spin-rotation experiments. The connection between the models and "Experimental Realities" is expressed clearly in the following paragraph:

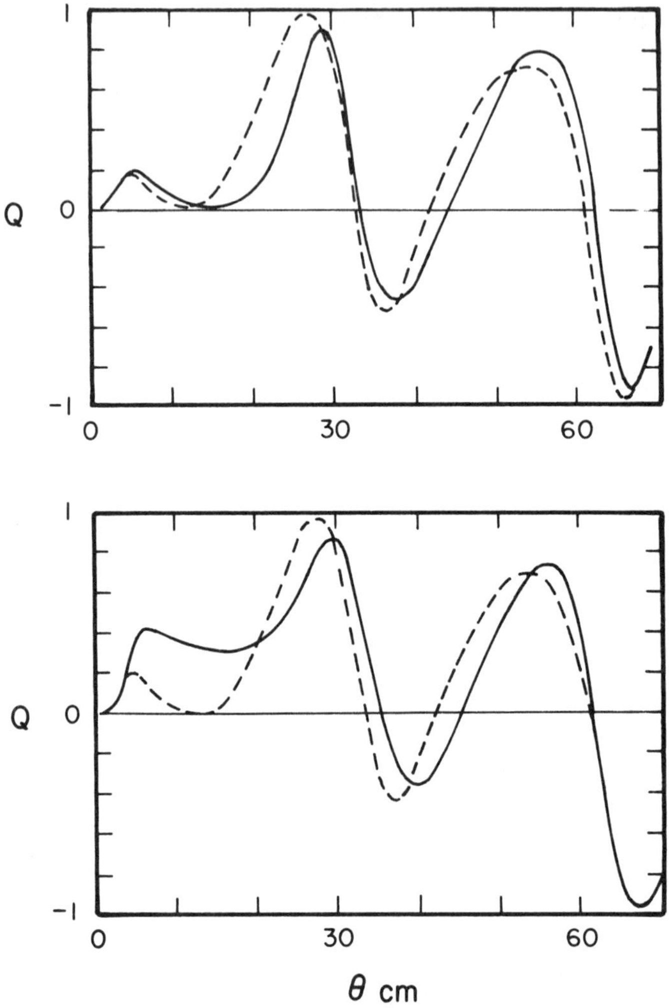

Figure 6. Predicted values of the spin rotation parameter, Q, for two different Schroedinger models (top) and two different Dirac models (bottom).

> We expect the statistical and systematic contributions to our determination of Q to be less than 0.03, a number substantially smaller than the range of variation among the predictions of these various . . . models. Thus these models will be substantially more constrained by the new measurements, and *discriminating choices among them may become possible*. (emphasis added)

In other words, the expected experimental error in measurements of Q is plus or minus three percent, but the differences between the values of Q predicted by the various models are substantially greater. So "discriminating choices" may be possible.

It requires almost no interpretation to fit this experimental design into a satisficing account of experimental tests. If one of the Dirac family of models is correct, the data points are very *likely* fall within three percent of the corresponding predictions for Q and highly *unlikely* to fall within three percent of the curve for Q predicted by Schroedinger models. And similarly if a Schroedinger model is correct. Thus, if one adopts the obvious decision rule (Choose Dirac if the data fits the Dirac prediction; Choose Schroedinger if the data fits the Schroedinger prediction.) one has a basis for a "discriminating choice." That is, one has a satisfactory decision problem.

Missing from the quoted description of this test is any mention of values other than the epistemic value of being right. This is as typical of research proposals as it is of published research papers. They are impersonal and cautiously understated—"discriminating choices *may* be possible." This is the only hint that there may be other factors operative in the choice to be made. But the caution need not represent value differences. It may just be a recognition of the possibility that neither family of models is very good and that the data might come out not clearly agreeing with the predictions of any of them. Indeed, this latter possibility is recognized in a standard joke among experimentalists. What is Murphy's law for experimental physics? The data you get is always the least informative possible.

The proposed experiment was approved and actually run in March, 1984. On about the fourth day of a six day run I found some members of the group, as usual, around a table by the computer in the data acquisition area. The data from the detectors goes directly into a computer and onto a tape. But there is a small subroutine that gives a quick and dirty calculation for Q just so the group can maintain some real time control on what they are doing. The experimental values of Q obtained so far were plotted by hand on standard graph paper. Scattered over the table were a half dozen or so graphs of predicted values Q for various models, some relativistic, some not. The experimentally obtained values were compared with the various predictions by superimposing one graph over the other. Indeed, there is a light table in one corner of the room to facilitate such inspections. Two of the predicted curves, one from a Schroedinger model, one from a Dirac model, are shown in Figure 7. Also shown are the data points collected up to that time.

A satisficing account of an experimental test would predict that this group would be most interested in measuring values of Q at scattering angles between 5 and 20 degrees. That is where the differences between the Dirac and Schroedinger predictions are generally the greatest. But that is not what we see here. There is one measurement at 10 degrees, but a half dozen at angles greater than 25 degrees. What is wrong?

In studying the actions of scientists there is no substitute for being able to ask informed questions. Why had they by this time not taken more measurements in the 5 to 20 degree range? In fact there are several relevant factors.

Most important, measurements inside 20 degrees are much more difficult than those outside 20 degrees because at the smaller angles the detectors are very close to the path of the primary beam. The experiment is designed to detect protons elastically scattered at a given angle. But the beam hitting the target produces all sorts of other reactions, including *inelastic* reactions resulting in the ejection of protons and other forms of radiation. This extraneous radiation is most pronounced in the forward direction, that is, along the path of the primary beam. At small angles, therefore, the detectors are being bombarded with all this extraneous radiation. By means of an elaborate coincidence and anti-coincidence sequence, the detecting system attempts to sort out the desired elastically scattered protons from everything else. However, if there is too much other stuff hit-

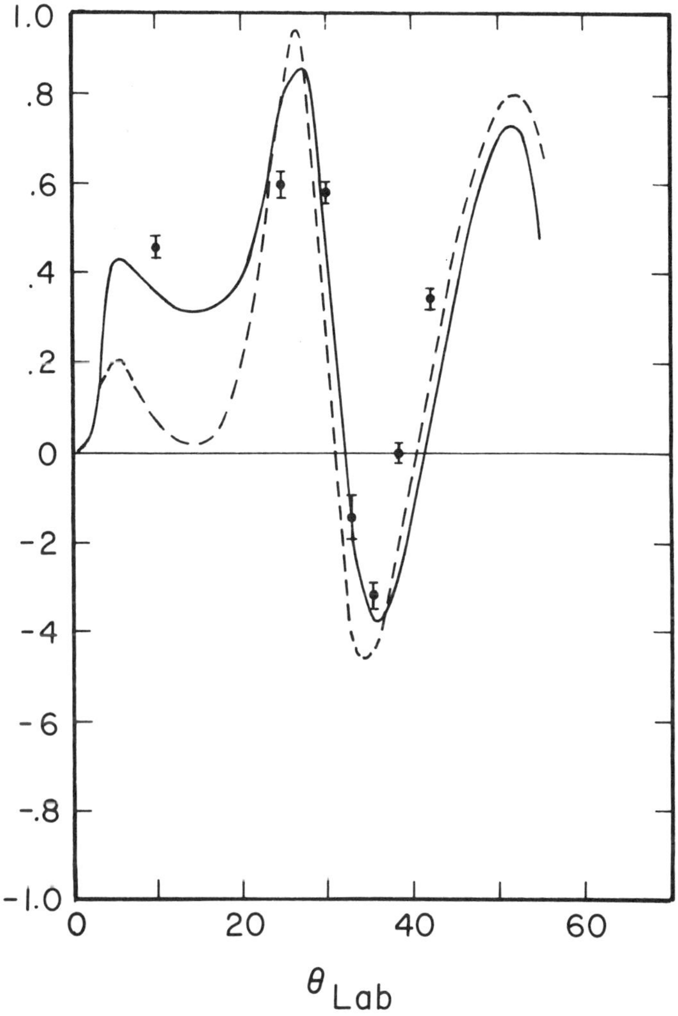

Figure 7. Comparison of one Schroedinger model and one Dirac model with preliminary data points.

ting the detectors, these electronic sorting devices get overwhelmed. They can no longer handle all the junk coming in and misrecord the desired events.

There are various ways of coping with his problem, but in the end the only way significantly to reduce overloading is to lower the intensity of the incoming beam. This means there will simply be fewer protons interacting with the target and less "noise." The trouble with this solution, of course, is that there are then fewer of the desired reactions as well. Thus the time required to get enough of the desired events to keep the statistical margin of error within the required bounds goes up. So the time per data point increases.

A similar problem arises at very large scattering angles, say beyond 50 degrees. For any incoming proton, the probability of being scattered at a larger angles is less probable than at a smaller angle—just as for billiard balls. So the counting rate decreases for measurements at larger angles, again increasing the run time per data point. As far as ease of taking data is concerned, the preferred angles are in the range between 25 and 50 degrees.

There is a second factor influencing the decision to take the first data at these experimentally easiest scattering angles. This run was the first with all the newly constructed equipment in place. It was decided to do the relatively faster and easier data points first in order to build confidence with this new equipment. The more difficult points were saved until after the team could be confident that the equipment was working as intended.

In addition, measurements at the smallest possible angles, down to roughly 6 degrees, require a change in the physical configuration of the reaction chamber and detectors. This means breaking the vacuum on the reaction chamber, making the required changes, and then reestablishing vacuum before data taking can resume. This takes perhaps four hours, which is a lot when one only has six days to complete all data taking. This fact means that one should probably save that data point until last so that one does not have to waste time getting things back into the more normal configuration. But this requires a careful judgment as to how long it will take to acquire that data. In particular, one must take into account the possibility that something may happen that would interfere with data taking—like a breakdown in the cyclotron itself.

So once one understands the more detailed circumstances of the experiment, one sees that this group of experimentalists was in fact doing just what a satisficing model of experimental tests would predict. They did regard the points in the 5 to 25 degree region as the most important and the organization of data taking was directed at obtaining the most reliable possible values of Q at just these angles.

Finally, note that even while working on the easier angles, the group seemed to be following a satisficing model. They did not just measure values of Q at regular intervals, say every 5 degrees. Rather, the first six data points were taken at scattering angles of 49, 42, 38, 36, 33, and 31 degrees. What determined this irregular pattern? Obviously the fact that one of the distinctive differences in the predicted curves for Q is the shape and location of the second minimum out around 35 degrees. They were trying experimentally to distinguish the models in just the way a satisficing strategy would predict.

XI. Why the Question Is Not Yet Resolved

Which approach, the Schroedinger or the Dirac, was favored by the experiment just described? Science is not so simple as the question implies. What I have described was just one of several "runs" of this experiment that took place over the next year. By the time the data was fully analyzed, the theoretical situation had also changed. More complex models of both types were developed by considering potentials that previously had been ignored as insignificant. When the values of the spin rotation function were computed for these more complex models, there remained little difference between the predictions of the two types of models. Some members of each family of models were pretty well ruled out, to be sure, but others of each type remained in the game.

Figure 8 shows a set of spin rotation predictions for two Schroedinger models and one Dirac model—together with the existing data points from the set of experiments done by the Indiana University team. Their judgment was that the data provided no solid basis for choosing one model over the others. And in fact the issue is

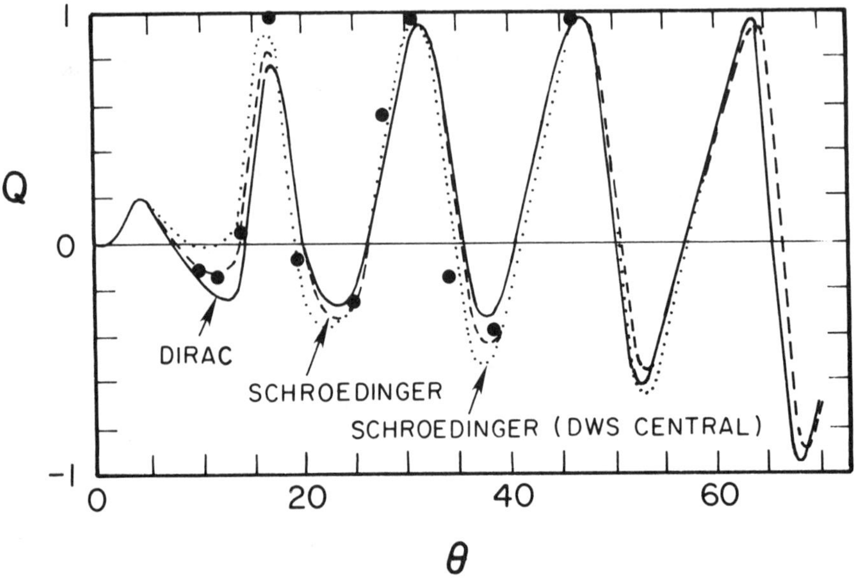

Figure 8. Comparison of final analyzed data with one Dirac model and two different Schroedinger models.

still (Spring, 1986) unresolved among members of the wider community of nuclear physicists concerned with such questions.

A satisficing analysis of the situation explains this lack of any clear decision. The data were judged to be roughly *equally* probable on either type of model. There is thus no available decision rule keyed on the data that provides a high probability of achieving one of the two satisfactory outcomes. This means that there is no satisfactory option available to an open minded satisficer. So no decision is made.

Philosophers of science, whether of an analytical or historical persuasion, are apt to be uncomfortable with this conclusion. In analytical terms, one of the two types of models should be rationally preferrable, if not, in fact, more probable. From a more historical standpoint, one of the two research traditions, the Schroedinger or the Dirac, ought to be the more progressive. It is up to philosophers of science to determine the criteria of rational preference or progressiveness, and apply them to this case. My view is that no such

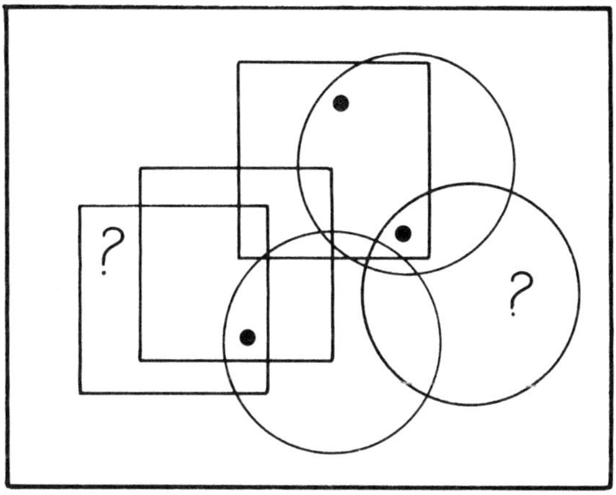

Figure 9. Representation of competition between Schroedinger models (open squares) and Dirac models (open circles). The solid dots represent experimentally determined phenomena.

criteria exist. If there is no satisfactory option available to an individual scientist, regarded as an open minded satisficer, then no decision will be made.

Figure 9 is an attempt to capture the situation graphically. The background represents the field of nuclear phenomena that are in principle detectable—whether they are detectable now or indeed have ever been thought of. The solid dots represent phenomena that have been reliably determined, such as the spin rotation parameter for 200 *MeV* protons on a Ca^{40} target. The open squares represent members of the Schroedinger family of models and the open circles members of the Dirac family.

The current situation is that all the known experimentally determined parameters are pretty well covered by both families of models. What would be required for a satisfactory decision between the two types of models is some experimental data on phenomena outside the reach of one or the other of the two types of models. Such possible phenomena are represented in Figure 9 by question marks. Such phenomena would be very probable on one type of model but highly improbable on the other. That would make possible a satisfactory decision rule.

XII. Conclusion

This episode in recent nuclear physics does not, by my own account, provide a decision test of a cognitive approach to the study of science. At best it illustrates the potential fruitfulness of such an approach and provides some basis for deciding that it is worth pursuing. And it gives some hope that the current bifurcation of analytical and historical approaches in the philosophy of science can at last be overcome.[12]

NOTES

1. In reflecting on these institutional developments one should not forget the role of Sputnik in stimulating interest in, and funding for, all things connected with science during the late 1950s and early 1960s.

2. For a much more extended discussion, complete with an account of many scientific results, see Ruse (1986).

3. For an already classic study of vision from a cognitive point of view see Marr (1982).

4. The notion of "schemata" goes back almost a half century to the work of Piaget (Gruber and Voneche, 1977) and Bartlett (1932). E. C. Tolman (1948) talked about "cognitive maps" in both animals and humans. For a more recent discussion, see Chapter 6 of Neisser (1976). Johnson-Laird (1983) has popularized the term "mental model" within cognitive psychology while Minsky (1975) introduced the term "frame" into the computer science literature.

5. See Nisbett and Ross (1980) and Kahneman, Slovic, and Tversky (1982).

6. For a relatively "popular" discussion of these developments see the review article, "Relativistic Treatment of Low-Energy Nuclear Phenomena," *Physics Today*, vol. 37 (1984), pp. 20–22. A much more technical account is given in Hoffmann (1985).

7. Kahneman, Slovic, and Tversky (1982), pp. 156–57.

8. See Simon (1945), Part IV of Simon (1957), and Parts II and VIII of Simon (1983).

9. Hempel (1960) introduced the notion of "epistemic values" into the recent literature. It has been used extensively by Levi (1967, 1980).

10. See Nisbett and Ross (1980) and Faust (1984).

11. I have borrowed the idea of open mindedness from Shimony (1970) who introduced it as part of his particular brand of Bayesianism called "tempered personalism."

12. The support of the National Science Foundation through its program of research grants in History and Philosophy of Science is hereby gratefully acknowledged.

REFERENCES

Bartlett, F. C. (1932) *Remembering* (Cambridge, England: Cambridge University Press).

Campbell, D. T. (1974) "Evolutionary Epistemology." In P. A. Schilpp, ed., *The Philosophy of Karl Popper* (La Salle: Open Court), pp. 413–463.

Faust, David. (1984) *The Limits of Scientific Reasoning* (Minneapolis: University of Minnesota Press).

Gruber, H. and J. J. Voneche, eds. (1977) *The Essential Piaget*. (New York, Basic Books).

Hempel, C. G. (1960) "Inductive Inconsistencies," *Synthese* vol. 12, pp. 439–469.

Hoffmann, G. W. (1985) "The Role of Experiment in the Development of the Relativistic Impulse Approximation for Proton + Neucleus Elastic Scattering." In J. R. Shepard, C. Y. Cheung, and R. L. Boudrie, eds. *Proceedings of the LAMPF Workshop on Dirac Approaches to Nuclear Physics*. Los Alamos, NM: Los Alamos National Laboratory.

Johnson-Laird, P. N. (1983) *Mental Models*. (Cambridge, Mass.: Harvard University Press).

Kahneman, D., P. Slovic, and A. Tversky. (1982) *Judgment Under Uncertainty: Heuristics and Biases* (Cambridge: Cambridge University Press).

Lakatos, Imre (1970) "Falsification and the Methodology of Scientific Research Programmes," in I. Lakatos and A. Musgrave, eds. *Criticism and the Growth of Knowledge* (Cambridge: Cambridge University Press).

Laudan, Larry (1977) *Progress and Its Problems* (Berkeley: University of California Press).

Levi, Isaac (1967) *Gambling with Truth* (New York: Alfred A. Knopf).

Levi, Isaac (1980) *The Enterprise of Knowledge* (Cambridge Mass.: MIT Press).

Marr, D. (1982) *Vision* (San Francisco: W. H. Freeman).

Minski, M. (1975) "A Framework for Representing Knowledge." In P. H. Winston, ed., *The Psychology of Computer Vision* (New York: McGraw-Hill), pp. 211–277.

Neisser, U. (1976) *Cognition and Reality* (New York: W. H. Freeman).

Nisbett, R. and L. Ross. (1980) *Human Inference: Strategies and Shortcomings of Social Judgment* (Englewood Cliffs, NJ: Prentice-Hall).

Putnam, Hilary (1982), "Why Reason Can't be Naturalized," *Synthese* vol. 52, pp. 3–23.

Ruse, Michael (1986) *Taking Darwin Seriously* (Dordrecht-Holland, Reidel).

Shimony, Abner (1970) "Scientific Inference." In R. G. Colodny, ed. *The Nature and Function of Scientific Theories* (Pittsburgh: University of Pittsburgh Press).

Simon, H. A. (1945) *Administrative Behavior* (New York: The Free Press).

Simon, H. A. (1957) *Models of Man* (New York: John Wiley & Sons).

Simon, H. A. (1977) *Models of Discovery* (Dordrecht-Holland: D. Reidel).

Simon, H.A. (1983) *Models of Bounded Rationality*. 2 vols. (Cambridge, Mass.: M.I.T. Press).

Simon, H. A., P. Langley, G. Bradshaw and J. Zytkow. (1986) *Scientific Discovery* (Cambridge, Mass.: MIT Press).

Stegmueller, Wolfgang (1979) *The Structuralist View of Theories* (New York: Springer).

Tolman, E. C. (1948) "Cognitive Maps in Rats and Men." *Psychological Review* vol. 55, pp. 189–208.

Toulmin, Stephen (1972) *Human Knowledge* (Princeton: Princeton University Press).

van Fraassen, B. C. (1980) *The Scientific Image* (Oxford: Oxford University Press).

EXPLANATORY SUCCESS AND THE TRUTH OF THEORY

ERNAN MCMULLIN

When Aristotle, in his account of the nature of *epistēmē* in the *Posterior Analytics,* looked for examples to illustrate the notion of demonstration which he claimed to be the defining characteristic of science, he turned to astronomy. On the face of it, this would not seem surprising, since astronomy might plausibly have been regarded as the most obvious candidate for a science of nature in his day. Yet the examples he chose were puzzling. Recall that demonstration had to proceed from premisses themselves known to be true; to "demonstrate" was to show how the observed properties of things proceeded necessarily from the essences of those things. One might have expected Aristotle to choose examples where these essences were familiar to us so that the truth of the premisses could readily be guaranteed.

Instead, however, he drew his examples from the sky. Why does the moon have phases, crescent, half, and so on? Because it is spherical. Why do the planets (unlike the stars) shine steadily? Because they are relatively near. To "demonstrate" waxing and waning as a property of the moon was to show how it followed necessarily from the spherical shape of the moon (and its being illuminated from afar by the sun). To "demonstrate" that planets would shine with a steady light, it was sufficient to show that nearby objects like them would shine in this way. But how were we to *know* that the moon is spherical or that the planets are relatively near? We would have to *know* this to be true if the premiss were to function as the starting-point of demonstration.

Aristotle's account is elliptical at this point[1] What he seems to suggest is that what assures us of the truth of the claim that the moon is spherical is that this does indeed explain the observed appearances of waxing and waning. We do not observe the shape of the moon directly, anymore than we do the nearness of the planets. But what allows us to take these properties as truly characterizing ones is that doing so enables us to explain the immediately observed properties. Explanatory power here serves as the criterion of truth in cases where the truth cannot be more directly ascertained. What is striking, once again, about Aristotle's choice of examples is that he deliberately chooses cases where the truth of the premisses could not be plausibly supposed to be known in advance. (One might have expected him instead to begin from a premiss like: man is rational; such an attribution on our part would not need to depend for its truth upon the success of the explanation this provided of other features of human behavior.)

Aristotle's account is problematic, of course, and later generations would struggle to make a consistent theory of demonstration out of it. I begin with it here in order to underline that the relation between explanatory power and truth has been a central issue in the understanding of science right from the beginning. In Aristotle's day, the hard cases were the physical properties of distant objects not open to immediate inspection, and therefore to be known only indirectly. In recent centuries, the objects of scientific interest have become far more remote, not so much in distance as in their relation to our intuitive understanding, the sort of understanding on which Aristotle's conception of science depended. The relation between explanatory power and truth, never an entirely easy one to explicate has thus become far more problematic.

I. Theory, Explanation, Truth

What I want to argue here, in the context of theory as we know it in natural science today, is that the only means open to us of judging the *truth* of a theory is through an assessment of the explanatory success of the theory. More important, this assessment *does* allow us, in favorable cases, to make a truth claim of a limited sort for the theory. To make my case, I should lay out very briefly first some familiar distinctions in regard to the three key terms in my thesis: "theory," "explanatory success," and "truth." The first of these may best be approached by contrasting two types of general statement found in science. A "law" derives from an observed regularity, and is arrived at by a process of generalization or induction. The inference from observed singulars to such a law is ampliative, that is, it goes beyond the evidence. But it does so only in the sense that it takes the instances observed to be a fair sample of a broader class. And of course the adequacy of the language used is also presupposed. A law-statement, in short, postulates a correlation between observable factors, on the basis of some limited experience of such a correlation.

A theory, on the other hand, is proposed as tentative causal explanation of a law or laws, as a way of explaining why the law holds. The joint process of arriving at theory and validating it is called "retroduction" by Peirce, because it means moving backward in thought from an observed effect to a hypothetical cause. It is ampliative in a much stronger way than is induction since it involves some degree of invention and imagination. One has initially to guess at the cause. The concepts in which the effect is described will not ordinarily suffice to describe the cause, and so an innovation of language may be needed. There are no rules (as there are in the case of deduction) that would tell us whether we are making the inference successfully.[2] In fact, even the term "inference" itself is misleading, since it might suggest something rule-governed. Conjecturing a cause, or more broadly, postulating a theory, is not rule-governed, although it will be influenced, up to a point, by prior commitments as to what counts as plausible.

What is important to us here is that theory cannot be assessed, as deductive inferences are, by the extent to which certain rules have been followed in its formulation. The criteria employed in theory evaluation are quite complex, and do not depend on how the theory has been arrived at. They have been attracting a good deal of attention of late since the point was first made that they function as values rather than as rules.[3] Here are just a few of the more obvious ones.[4] First, of course, comes empirical adequacy (or predictive accuracy), that is, the ability of the theory to account for the data at hand in a deductive (or at least a quasi-deductive) way. A second is proven fertility, the ability that the theory has already demonstrated of predicting "novel" facts, as well as the ability it has shown of dealing with anomaly by imaginative modifications suggested by the theory itself.[5] A third is unifying power or scope, the ability the theory has shown of bringing together domains hitherto taken to be disparate. A fourth would be consistency with background presuppositions and other accepted theories. Fifth is coherence, "naturalness," the absence of those *ad hoc* features that can so damage an hypothesis, even one that otherwise "works." More "internal" than the others, finally, is logical consistency. Each of these considerations would obviously require more detailed discussion, but this summary listing may suffice for our purposes here.

These are some of the "values" scientists look for in a theory, considered as an explanatory hypothesis. Taken together, they allow an estimate to be made of the *explanatory success* of the theory. A failure in any one of them is regarded as a failure in explanatory power. Empiricist philosophies of science have tended to assimilate explanation with prediction, and hence to focus on the first criterion only, as though one could evaluate a theory simply in terms of its predictive accuracy. This was especially appealing to the logical positivists because it seemed obvious to them that the warrant for a theory could lie only in its empirically verified logical consequences. The inductive strength of an hypothesis could come, they argued, only from its deductive relationships with the observational evidence. Other considerations might, indeed, intervene but they could be regarded as "aesthetic" or "pragmatic"; they were not to be counted towards the *truth* of the hypothesis. Nor could they, in consequence, be included in the requirements for explanation proper.

The difficulty with this approach was, of course, that the primary function of a theory is to explain, and that its failings as explanation would seem on the face of it to count against its credentials as a good theory. A theory with features that are perceived as *ad hoc*, for example, may be perfectly adequate as a means of accurate prediction, but these features will count against the theory as explanation and will prompt efforts to find an alternative. One could say that this is because it is anticipated that such a theory will ultimately fail as a means of prediction because of the presence of these *ad hoc* features, so that predictive accuracy might still be regarded as in some sense the primary value. But the fact remains that the evaluation of a theory does *not* simply reduce to asking about its present efficacy as a predictive device.

These criteria have been described above as a sort of "given," and of course this is much too simple. Have they changed over time? On what does their warrant *as* criteria of theory rest? On the consensus of practicing scientists? On their effectiveness as leading to predictively powerful theories? On their intrinsic logical merits? These issues have given rise to lively discussion among philosophers of science in recent years, and no resolution of the differences appears to be in sight. It is fair to say that these values *are*, in fact, employed in theory-evaluation in contemporary natural science. There would not, however, be agreement among working scientists as to precisely how they should be understood or how much weight to attach to each. It is, indeed, this lack of agreement that explains in great part (though not entirely) the persistent presence of controversy in natural science, something that on the older simpler accounts of the theory-evidence relationship was quite hard to understand.

For the moment, I want to set aside questions about the status of these epistemic values and simply to note that there would, in the case of longer-standing theories at least, be a considerable measure of agreement among working scientists as to the degree of explanatory success to attribute to the theories. How does this success bear on the *truth* of the theory? Much, of course, depends on what one means by "truth" in this context. In the period between the two world wars, there was a great deal of discussion of the different "theories" of truth, prompted in the first instance by the writings of Moore and Russell. Philosophers debated about correspondence and coherence and pragmatic success, and generated arguments in support of the claim that truth resided in one and only one of these. I shall return to this debate later, but for the moment let me recall Aristotle's simple dictum: "To say of what is that it is, is true; to say of what is that it is not, is false." It is easy enough to apply this to "the cat is on the mat," or "snow is white." But what about a scientific theory? This is a very complex sort of statement indeed. What would it mean to say of such a theory that it is "true"?

A theory explains in the first instance by postulating a hypothetical structure of some sort, a set of entities, processes and relations that are causally sufficient to account for the observational data.[6] An example will help. Most of the major features of the earth's surface, such as the great mountain ranges, the undersea ridges, the volcanoes, as well as some less obvious features like the geological match between parts of the coasts of West Africa and eastern South America, and parallel magnetic striations on the ocean floor, can be explained by postulating the existence of immense plates, carrying continents and sea-floor alike, slowly moving relatively to one another, sometimes pulling apart and sometimes pressing together with force enough to thrust mountains into the sky.

To say of such a theory that it is *true* would seem most simply to be construed as saying of these plates "that they are." Or as Wilfrid Sellars puts it: "To have good reason for espousing a theory is *ipso facto* to have good reason for saying that the entities postulated by the theory really exist."[7] A theory in this reading is interpreted as a complex causal hypothesis, which is true if and only if the entities postulated in the theory actually exist and operate as they are supposed to do. The idea is that if the causal mechanism does *not* exist, the explanation does not hold, not as an explanation, at least.

But is there any other way a theory *might* hold? A much weaker thesis would be that a theory is true if its observable consequences, its predictions, all hold good. In this reading, the causal mechanisms postulated by the theory are assumed to be no more than heuristic devices whose function is to aid scientists in extending the range of their predictions. No existential claim is intended, and thus the truth of the theory does not require that anything more exist than the effects themselves.

But is "truth" the proper term to use in such a case? It would seem not, and indeed most of those who interpret theory in this way would reject the term entirely in this context. The classic instrumentalist view of theory was that it served as nothing more than an instrument of prediction. A *good* theory is one that gets all the predictions right. Such a theory is a tool, not a truth-claim, and thus the term "true" is not to be used of it at all. Bas van Fraassen proposes a view he calls "constructive empiricism" in which a theory is held to nothing more than its consequences.[8] A theory whose consequences are duly verified is said, not to be true, but to be "empirically adequate." In van Fraassen's view, no more may be asked of a theory than this. In particular, its success as a theory cannot be held to warrant in any way the stronger claim that the entities postulated by it exist, that is, that it is true.

I shall argue the opposite thesis, namely, that the success of a theory as a theory suggests the *truth* of the theory. And I mean this last phrase to be taken in its strong Sellarsian sense. Notice, though, that I only said "*suggests*." I am not saying that explanatory success *implies* the truth of the theory. The rest of what I have to say will be a gloss on this.

II. Scientific realism

At this point, it would be helpful to ask what the relationship is of all this to the doctrine known as scientific realism. In point of fact, there seem to be almost as many scientific realisms as there are scientific realists. I will have to rely here on my own definition of that much-fought-over label.[9] Scientific realism, let me stipulate, is the view that there is good reason to believe in the existence of entities substantially like those postulated by theories that have been successful over the long term. There are several important qualifications built into this definition. Briefly, I take realism to commit one only to saying that there is "good reason" and not that there are compelling grounds. It is always *possible* for even a highly successful theory, to be false; it may be extremely unlikely but we have to hold the logical possibility open. And the theory may develop further, so that the present specifications of the causal mechanisms may have to be revised and sharpened. Finally, only theories that have already shown a considerable degree of explanatory power would qualify as having reliable ontological implications. Each of these three reservations would need to be specified in further detail but I will have to leave that task aside here.[10]

How, then, does scientific realism so defined relate to the question of whether scientific theories may be said to be *true*? And, even more to the point, how do both issues tie in with the notion of explanatory success? My argument may be reduced to two constituent claims: first, that a scientific realist is committed to attributing *some* sort of truth-value to successful theories, and second, that the measure of truth-likeness or likelihood to be attributed to the theory is precisely the degree of explanatory success the theory is taken to have shown.

The first of these claims is easier to fill out than the second. The realist asserts that when a theory satisfies to a high degree the criteria just listed, this gives us good reason to believe in the existence of something like the causal mechanisms it postulates. But it is, in fact, the existence of these mechanisms that would make the theory true. Can we say, then, that a realist should hold that one can ascribe truth to a successful theory? This is indeed the way realists sometimes talk, but unfortunately the inference is a little too quick. If the theory really *is* a theory and not a covert description (such as the "theory" that the blood is circulated by the heart, a theory in Harvey's time, but no longer in ours), it is inferring to causal mechanisms that are themselves not directly accessible to us, at the moment, at least. The manner in which it does this, the concepts in terms of which these mechanisms are described, in short, the theory itself, is subject in principle to further development. Something left blank in the present model may get filled in. The concepts in terms of which the theoretical entities are specified may be modified, perhaps even considerably modified.

The success of the plate-tectonic model in the two decades since it first gained wide acceptance among geologists around 1965 has been striking in every regard. It has, for example, managed to explain whole ranges of data that were not originally even seen to be related to it. Would this lead us to say that the theory is *true*? Clearly not. It is still under development. We can be quite sure that there are still anomalies ahead that will lead to modifications in the original model, just as the plate-tectonic model itself involved a fundamental change in the continental drift model that prepared the way for it. If by "true," we mean "literally true," that is that the causal mechanisms are *exactly* as the present theory postulates them to be, then we can be fairly sure that the theory is *not* true. To say that it is true would be to assert that nothing further will be discovered in all the wide domain of geology that would be sufficiently anomalous, relative to the present model, to lead to a modification in this model. It does not require very much familiarity with the history of science to know that even the highest degree of explanatory success in the case of a large-scale theory would not warrant such confidence.

A successful theory ought not, therefore, be described without qualification as "true." Aristotle long ago required science to generate eternal and necessary truth. We would be happy to settle for just plain truth. But even that is beyond our powers in all but the most limited explanatory contexts in natural science. Scientists recognize this fact and ordinarily are quite careful not to describe even their most cherished theories as "true," preferring weaker terms like "reliable" or "well-supported." And philosophers are similarly cautious, proposing substitutions like "warranted assertibility" for the older and more demanding term.

Someone who accepts a version of scientific realism would seem justified, however, in employing somewhat stronger language. A realist ordinarily allows that a high degree of explanatory success on the part of a theory gives good grounds for asserting the existence of entities of the general kind postulated by the theory. But this is sufficient to allow one to say that there are good grounds for asserting the theory to be "approximately true" or "highly likely." Critics have objected to the term "approximate" in this context, both because the term is vague and because the degree of approximation cannot, in principle, be specified. One cannot know in advance how much and in what directions a scientific theory may still have to be modified.

While this is true, there is nevertheless a sense of "approximate truth" that would seem perfectly defensible in this context. Consider again our geological example. The striking success of the plate-tectonic model in so many domains gives geologists good reason to believe it to be *true* that the continents and ocean floors are carried on massive rocky plates in slow relative motion. There is much about these plates and about the mechanisms repsonsible for their motion that geologists can as yet only guess at. In that sense, the theory is incomplete, open to further refinement. But suppose we focus not upon the specifics of the theory but upon the very general claim that there are large plates in relative motion. Geologists would, I think, say that we have very good grounds for believing *this* claim to be true.

Calling a theory "approximately true," then, would be a way of saying that entities of the general kind postulated by the theory exist. It is "approximate" because the theory is not definitive as an explanation; more has yet to be said. But it already has a bearing on truth because we can say that it has allowed us discover that entities of a certain sort exist, entities that we could not (for the moment, at least) have known about without the aid of the theory.

To the extent that a theory postulating quite different sorts of explanatory entities is still an open possibility, we would want to qualify the claim made, and say that the original theory is "probably" approximately true, or less cumbrously, "highly likely." The notion of likelihood is useful here because it implicitly contains both types of qualification on the truth-claim being made, first that it is only "approximate" in the sense specified above, and second that there is a chance, though perhaps only a very small one, that entities of the general kind postulated do not exist after all.

Some other examples may help to fix the kind of claim that I am making. Astrophysics has developed steadily since the seventeenth century. At this point, we can be reasonably assured that the stars are large glowing masses of gas, containing chemical elements similar to those of earth. There is such an abundance of evidence converging on this general claim that it would be given at least as much credence as would the average report of a laboratory result. Much more than this can, of course, be said about the composition of individual stars, about the sources of their energy, about their likely past evolution and future development, about their grouping at very different distances from us, and so on. The relative assurance with which each of these further specifications can be asserted will depend on the explanatory power of the associated model and theory. But for the moment, we are interested only in the most basic claim, namely that the points of light in the night sky are either individual globes of incandescent gas like our own sun, or groups of such suns.

Any other mature structural science will serve to make the same point. The "billiard-ball" atom of the kinetic theory of gases, for example, has gradually taken on a more and more detailed structure. The existence of atoms themselves as structured individual entities of known mass and volume can now be asserted with assurance, even though much has still to be learned about the details of their structure. Our atomic theories are unlikely to remain in precisely their present form. But this does not prevent us from making a more general claim about atoms, basing it on the widest possible appeal to the explanatory success of a cluster of associated theories.[11]

III. Realism without Theory-Truth?

I have been arguing that scientific realists are committed to a truth-claim of *some* kind about the theories in which the entities whose existence they assert make their appearance. But would any realist actually *deny* this? Some would, at least, have doubts. In his recent book, *Representing and Intervening*,[12] Ian Hacking defends a form of realism that relies entirely on an experimental tracing of effect back to cause, avoiding any commitment to the truth-status of theories. What he asserts is not the existence of theoretical entities but of "phenomena" that can be produced experimentally. Since the photo-electric effect was first discovered in 1829 by Becquerel, he reminds us, one theory after another has been put forward to explain it. The current (photon) theory may be rejected in time to come, but "the supermarket doors (which depend on the photo-electric effect for their functioning) will still go on working."[13] Phenomena, once established, are not affected by theory-change. "The phenomena which we have created will still exist and the inventions will work." Once phenomena like the photo-electric effect are created in the laboratory, "they are phenomena thereafter, regardless of what happens." He calls this "experimental realism."

But who would disagree with this? Not the instrumentalist surely? Hacking is willing, it is true, to go beyond such "phenomena" to entities like electrons that "in principle cannot be 'observed'," as long as they can be experimentally "manipulated to produce a new phenomenon."[14] We come to know the mass and the charge of electrons and enough of their "causal powers" to enable us to "build devices that achieve well-understood effects in other parts of nature." Not only does a *theory* of electrons play no part in ensuring the reality of these experimental entities, it does not even serve to define or describe them either. There is, in fact, "no common core of theory," only a "a lot of theories, models, approximations, pictures . . . ."[15]

It is important to realize that Hacking opposes his experimental realism not to instrumentalism but to *idealism*. The theme of experimental intervention contrasts "doing" with "thinking," manipulation with explanation. In this context, to argue that something is "real" is to argue that it is not simply a creation of mind, that it imposes itself *on* mind. The category he perceives as threatened is not the unobserved entity but the "phenomena" themselves; the instrumentalist would find no difficulty in sharing this concern.

His second target is the extreme form of empiricism that leads van Fraassen to deny that we can "observe" through a microscope and to reject, in consequence, the reality of entities attested to only by such instruments. Hacking uses his notion of intervening once again to argue quite persuasively that the entities viewed in the microscope can be experimentally tested for "reality" in numerous ways, that is, shown not to be artifacts of the experimental techniques employed. The truth of the reality-claim in their regard does not depend in any way (Hacking argues) on having a *theory* about them. We may have no understanding at all of the organism we are observing under the high-power microscope, yet we might still be quite sure that it is real.

This example brings out both the strength and the limitation of Hacking's argument. The entities "observed" by scientists are ordinarily identified by a term, like "electron," that carries with it a "theory" of the entity involved. When a scientist claims to have observed an electron in a cloud-chamber, there is not only a causal tracing but an implicit commitment to the theory that enabled this identification to be made in the first place. If the theory were to be rejected, scientists would still suppose they had observed *something* but they would no longer report this observation in the same way. Calling the object an "electron" enables them to relate it to other entities in a network of relationships prescribed by theory. The case of the investigator who notes, for example, a hitherto unidentified reddish-colored needle-like entity on his microscope slide and suspects it to be (say) the agent that caused Legionnaire's disease is untypical. Once, however, this entity is classified as a virus or as a bacterium, a whole set of theoretical expectations immediately come into play.

What Hacking is telling us is that our reason for believing in the existence of the unidentified organism has nothing to do with a *theory* of that organism. Even in the case of the electron, our reason for believing that the electron exists is primarily the causal tracing. The theory is needed only for identification's sake and not for the assurance of existence. The weight of an existence-claim must rest on intervening, he insists, even though an account of the outcome of the intervening will surely require some degree of representing.

Like Kuhn, he finds theory too weak a reed to support reality claims. But his causal argument for the existence of such an entity as an electron will hold (he reminds us) even if the *theory* of the electron should change drastically. Of course, the meaning of the existence-claim will change in such a case, but Hacking is not worried about this. Intervening makes it possible to know *that* the electron is, even though there is no similar assurance as to *what* it is. But once again, unless the "what" can be identified—and how else except in terms of theory?—the "that" would seem to be empty. It is not clear, then, that reference to explanatory success can be dispensed with, even in an argument that relies mainly on experimental intervention.

Hacking's argument for realism leaves aside most of the entities and processes that the theoretical sciences deal with. When chemists postulate hydrogen bonds between the component atoms of a molecule, for example, it is not on the grounds that these are "observed," even in the extended sense of that term advocated by Hacking.[16] The warrant for entities such as these is the success of the theories of which they form a part, nothing else. And what about objects distant from us in space (e.g. galaxies), or in time (e.g. dinosaurs)? Causal processes link us to them, it is true, and in the former case (though not the latter) we might even say that they are "observed." But no "intervening" is possible, so they do not appear to qualify for support under Hacking's argument. It is, however, this profusion of theoretical structures that lie beyond instrumental intervention that gives science much of its interest. Hacking does not deny their existence; he merely expresses his doubts as to whether valid arguments can be found to support existence-claims on their behalf by appeals to explanatory power alone.[17]

One conclusion we can draw from all this is that a broadly "realist" approach to the explanatory entities of the scientist can be defended in two quite different ways. An "intervention" type of argument is applicable only to a sub-class of these entities; it has the advantage that it affords a relatively direct inference to existence. The argument from explanatory success is much more general in its application but is more vulnerable because of the complex ways in which theoretical concepts refer and the ever-present likelihood that the theories may change. This argument takes the same form whether or not the entities it postulates are "manipulable," in Hacking's sense, or not. The two arguments can, of course, be combined for "manipulable" theoretical entities, affording a greater degree of assurance in their regard. What the "intervention" type of argument does is to ensure that there *is* an entity there to begin with, even though it may have little to say about its nature other than that it causes certain effects. The argument from explanatory success goes on to fill in the nature of the entity, so that the two complement one another. One can see this dual process at work in the history of the electron concept, where from the beginning the electron was "whatever it is that is causally responsible for these experimentally-determined results," as well as a complex construct in a sequence of explanatory theories.

Where reference depends on theory alone, an argument from explanatory success is required in order to support the initial claim that an entity of this general sort exists. The familiar stories of phlogiston and caloric remind us that such assertions are defeasible. Note that the claim here is for an entity *of this general sort*, not of this *precise* sort. Theoretical entities are idealizations, possessing only the attributes given them by the theory. The success of the theory does not warrant the claim that something *exactly* corresponding to this construct exists; such an entity would have no "material" aspect, no further properties to explore.[18] The success of a theory can at best warrant one only in claiming that an entity possessing *among others* the properties attributed to it by the theory exists. And since the theory is unlikely to be definitive, it is more appropriate to say that certain properties the referent of the construct possesses are likely to be of the general sort prescribed by the theory.

The importance of these qualifications lies in the fact that theories develop and are modified, sometimes in fundamental ways. In this section I have been arguing that a realist is committed to attributing likelihood to successful theories in science. But realism does not amount to holding that entities exactly and exhaustively described by the constructs of successful theory exist. Such a thesis would be defeated by the smallest theory-change. The continuity of reference of successful constructs is a much looser affair, as case-studies have shown. Such constructs suggest certain articulations of Nature, but these articulations may alter as theories develop and converge. The Dirac electron is completely specified by Dirac's theory. But the *real* electron, the entity whose nature has gradually become clearer, thanks to a confluence of theories and a strong intervention-type argument, is not exhausted by any theory we now have or for that matter ever may have. These are notoriously complicated issues; limitations of space prevent a fuller treatment.

IV. Inference from best explanation

My second thesis is that the likelihood of a theory depends on the degree of explanatory success the theory enjoys, construing "explanatory success" here in the broadest possible way. This resembles what Harman has called "inference to the best explanation":

> In making this inference, one infers from the fact that a certain hypothesis would explain the evidence to the truth of the hypothesis. In general, there will be several hypotheses which might explain the evidence, so one must be able to reject all such alternative hypotheses before one is warranted in making the inference. Thus one infers, from the premise that a given hypothesis would provide a "better" explanation for the evidence than would any other hypothesis, to the conclusion that the given hypothesis is true.[19]

Harman's label is not an entirely happy one. If he were right in saying that one must be able to reject *all* alternative hypotheses (something that is, in principle, hardly ever possible in science) before the inference were warranted, then we would have inference not just to the best, but to the *only*, explanation. Furthermore, we do not *infer* to a hypothetical explanation, strictly speaking, we *conjecture* it. Where inference comes in is in concluding that it *is* the best available explanation. We do not infer to the best explanation; we infer that a given explanation *is* the best available explanation. In the passage quoted, Harman himself says, not that we infer to the best explanation, but *from* the claim that a given hypothesis is the best explanation *to* the conclusion that this hypothesis is true.

We have already seen that this last is far too strong. From the premise that an hypothesis is the best available explanation, we *cannot* infer that it is true, only at best that it is likely. Moving from best explanation to truth is the troublesome issue. Harman's problem is not with the inference that something is the best explanation (indeed, he dismisses this in a few lines), but rather with what one can infer *from* the assertion that a given hypothesis is the best explanation. Hence, inference *from* the best explanation (or even better, from explanatory success) would be a more appropriate label. The older term coined by Peirce, "retroduction," is in any event already available for this purpose.

How is inference from best explanation to the likelihood of the corresponding hypothesis to be supported? Earlier, we listed the three so-called "theories" of truth: correspondence, coherence, and pragmatic effectiveness. It has frequently been noted that the latter two are *criteria* of truth and are much less plausible as means of defining the nature of truth. Whereas correspondence quite obviously cannot be a criterion; it can at best only have something to do with the nature of truth.

Most of the difficulties that have been raised about truth-as-correspondence have been due to regarding it as a sort of *picturing* relationship. Wittgenstein and Rorty are only two of the many who have shown that the notion of picturing will not do in this context. But if this metaphor be set aside, one can retain the idea that it is the way the world is that determines what is true ("to say of what is that it is is true"). And the main criteria that allow us decide whether we have grasped the way the world is are clearly the coherence of the explanations we give and the accuracy of the predictions they enable us to make. These two generic criteria recall the criteria of theory-assessment already discussed. Coherence groups together a number of desirable values: internal consistency, consistency with other well-established theories, lack of *ad hoc* features, and so on. We also want a theory that is predictively accurate, whose practical applications work out as they should, and which enables a fruitful research program to be carried on. The criteria that scientists would regard as appropriate to the assessment of a theory resemble, therefore, to a striking extent the criteria traditionally thought to be symptoms of truth generally. The values that constitute a theory as "best explanation" are, it would seem, similar to those that would qualify a statement as "true."

I can put this in a more concrete way. Suppose, as the realist supposes, that a scientific theory may be said to warrant some sort of qualified truth-claim in regard to the entities constituting it as explanatory. How ought a theory so understood be evaluated? How, for example, would one assess the claim that the core of the earth is made up of molten iron? I submit that it would be (so far as we can tell) by applying exactly the criteria that scientists *do* apply to theory. These are very much the criteria one would expect scientists to use if theories are given a realist construal. They are *not* the criteria that an instrumentalist construal would lead one to anticipate. The coherence criteria, in particular, are realist in their implications.

V. The opposition

Why, then, would there be any objection to linking explanation and truth in the way that realists do? I will end by sketching very briefly the reasons why some prominent philosophers of science reject any such link, and add a few comments. van Fraassen, as we have already seen, takes the strongest line. "The belief involved in accepting a scientific theory is only that it 'saves the phenomena', that is, correctly describes what is observable."[20] Other criteria that affect theory-acceptance are purely pragmatic, and "pragmatic virtues do not give us any reason over and above the evidence of the empirical data for thinking that a theory is true." The aim of science is the construction of models with a view to prediction, "and not the discovery of truth concerning the unobservable." [21] His basic quarrel with realism is with its "inflationary metaphysics,"[22] against which he invokes the traditional "nominalist response."[23] Regularities in the world are brute fact; whether or not they can be explained in terms of unobservables, it "does not matter to the goodness of the theory, nor to our understanding of the world."[24]

There are some fundamental difficulties, I think, with the notion that the regularities in the world around us are brute fact; one remembers the sad tale of "grue," and how Goodman had to fall back on a notion of "entrenchment" to mark the boundary between lawlike and accidental generalizations. But what "entrenches" a generalization is its explainability in terms of a broader theory. This is surely a long way from brute fact. van Fraassen attempts a detailed nominalist response to this and other objections to his view, many of which were already formulated by critics of positivism.

Instead of pressing these objections, I will content myself with one very general counter-argument. Retroduction is constantly employed in everyday contexts, as well as in science, to infer to the existence of postulated causes of the effects to be explained. The existence of these causes is frequently verified directly at a later time thus (so far as one can see) justifying, to some extent at least, the confidence extended to the retroduction at an earlier stage. A favorite example of mine comes form the First Day of Galileo's *Dialogo*. Only the existence of lunar mountains, he argues in some detail, can explain the phenomena of changing shadows observed on the lunar surface. These mountains began, therefore, as theoretical entities, the fruit of inference from best explanation. Now that men have walked on them, their existence has presumably been established. Was belief in their existence unwarranted until the 1960's? Or was there not at least a partial warrant beforehand? The nominalist has difficulties in finding a consistent answer to the objection that direct observational tests very frequently bear out existence-claims originally derived from successful theoretical explanation. Does this not afford a reason to attach some credence to such claims even in the absence of a direct check? What is so special about direct observation that the likelihood of existence of mountains on the moon is negligible until the spaceships land? And if it is *not* negligible, it can have been based only on retroduction. The burden of proof is surely on the person who denies *any* connection between truth and explanatory power.[25]

Another philosopher who appears to deny this connection is Nancy Cartwright. In her recent book, *How the Laws of Physics Lie*, she argues that the explanatory power of scientific laws is gained at the expense of truth, so that in practice truth and explanatory success actually tend to *exclude* one another.[26] In arriving at this provocative conclusion, she makes a number of questionable assumptions about the nature of explanation in science. Though critical of the positivist *DN* model of explanation, she shares in at least some of its presuppositions. She assumes, for example, that the basic explanatory function in science is carried by laws, not by theories, and that the criterion of explanation is what she calls "organizing power." Explaining is thus a matter of subsuming under a lawlike generalization. The basic laws in physics are idealizations intended to "explain" in this sense, she goes on, but when they are applied to complicated real-world situations, many of them have to be amended in *ad hoc* ways. They are thus in the strict sense false; it is the attempt to give them a high degree of explanatory power that leads to this.

There are many difficulties, to my mind, in this argument. The basic explanatory statements in science are theories, not lawlike generalizations; retroduction rather than induction lies at the basis of scientific explanation. Furthermore, the techniques of idealization and of composition of causes do not, for the most part, falsify in the way she supposes.[27] Although there are instances where idealized laws have to be modified in *ad hoc* ways to make them fit concrete cases, these do not seem to be frequent, certainly not as frequent in physics generally as she assumes. And where such instances occur, the original law is regarded as defective even from the explanatory standpoint. The *ad hoc* character of its application to concrete cases—think of the renomalization techniques used in the quantum theory of the electron thirty years ago—count against it not only as truth, but also (in the scientist's eyes) as explanation.

The most serious challenge to the linking of explanatory power with truth may well be the Kuhnian one, based on the historical testimony of theory-change and theory-replacement. Kuhn comments on the "implausibility" of the claim that successive theories in science come closer to the truth or that truth lies in a match between theoretical entities and their counterparts in nature.[28] Laudan has developed this argument in some detail. He takes the realist to be linking truth and explanatory power in two respects, holding first that if a theory is approximately true, it will be explanatorily successful, and second that if it is explanatorily successful, it is probably approximately true.[29] He denies both of these conditionals, and argues that the notion of approximate truth is too vague in any case to carry weight in an argument of the sort the realist proposes. I have tried above to fill in this notion a little. It is worth recalling again that a *formal* measure of the degree of approximate truth of a successful theory is, in principle, impossible to specify.

Realists would not, I think, be committed to holding that if a theory is approximately true, it will *necessarily* be explanatorily successful. They might say that a theory is more *likely* to be explanatorily successful if it is approximately true than if it is false, that is, if nothing like the entities it proposes exists. But what the realist is more interested in is the claim that if a theory is explanatorily successful over a considerable time, it is probably approximately true. Laudan objects to this on historical grounds, adducing examples of successful theories which were later abandoned and thus cannot be held to have been approximately true.

My response to this objection would come in three stages, which cannot be developed here in the detail needed.[30] First, the realist insists on the qualification "probably," thus allowing for the possibility in some cases that a successful theory may be abandoned. Second, the historical instances Laudan cites are of very uneven value. In some cases (such as the crystalline spheres of medieval natural philosophers, or the theory of circular inertia) there was no long record of explanatory success in the sense in which this criterion was defined above. In others (such as the caloric theory), the earlier theory was reinterpreted, retaining much of the formalism. Each case would require separate discussion, and the value of Laudan's paper is that he has provoked such discussion.

But my main argument against Laudan's blanket rejection of the explanatory success-to-truth mode of argumentation is that it proves too much. Or as the medieval adage puts it: *Quod nimis probat, nihil probat*. (What proves too much proves nothing.) It is one thing to use the historical record to urge caution on anyone trying to formulate the realist claim precisely. But it is quite another to use it to reject the realist claim entirely, i.e. to maintain that explanatory success is *no* indication of truth. For this leaves us unable to make sense of the manifold instances when retroduction can be shown to have led to correct assessments of the truth. It prevents us from giving any sort of credence to existence-claims about even the best-supported theoretical entities. The historical record simply does not sustain so extreme a principle.

NOTES

1. A fuller discussion of Aristotle's enigmatic astronomical examples will be found in a preliminary version of this paper, "Truth and Explanatory Success", *Proceedings of the American Catholic Philosophical Association*, vol. 59 (1985), pp. 206-231.

2. Induction is an intermediate case. Scientists make use of guidelines (e.g. about curve-fitting) in the formulation of empirical generalizations, but the guidelines do not prescribe a unique form for the inferred law.

3. The point was, I think, first clearly made by Thomas Kuhn in his *Structure of Scientific Revolutions* (Chicago: University of Chicago Press, 1962).

4. For a fuller discussion, see my "Values in Science," in P.S. Asquith and T. Nickles (eds.), *PSA 1982* (E. Lansing: Philosophy of Science Association, 1983), pp. 3-25. See also Paul Thagard, "The Best Explanation: Criteria for Theory Choice," *The Journal of Philosophy*, vol. 75 (1978), pp. 76-92.

5. It is important to distinguish *proven* from *unproven* fertility as criteria. The former is a criterion of success, the latter a criterion of future promise. See my "The Fertility of Theory and the Unit for Appraisal in Science," in R. S. Cohen *et. al.* (eds.), *Boston Studies in the Philosophy of Science*, vol. 39 (Dordrecht: Reidel, 1976), pp. 395-432.

6. See my "Structural Explanation," *American Philosophical Quarterly*, vol. 15 (1978), pp. 139-147.

7. *Science, Perception and Reality* (New York: Humanities, 1962), p. 97. I would not agree with the further inference Sellars draws from this: "On the view I propose, the assertion that the micro-entities of physical theory really exist, goes hand in hand with the assertion that the macro-entities of the perceptible world do not really exist" (p. 96).

8. See *The Scientific Image* (Oxford: Clarendon Press, 1980), chapter 2.

9. In the introduction to his anthology, *Scientific Realism* (Berkeley: University of California Press, 1984), Jarrett Leplin lists ten theses associated with positions commonly called "scientific realism." He notes (p. 1) that no realist is likely to endorse a majority of them, even if they were subjected to reasonable qualification. And yet each of them can be found in the work of some realist.

10. I have developed this definition of scientific realism in some detail in "A Case for Scientific Realism," Leplin, *op. cit.*, pp. 8-40.

11. Wesley Salmon has developed this as an argument from a common cause; see "Why Ask: 'Why?'", *Proceedings of the American Philosophical Association*, vol. 51, (1978), pp. 683-705.

12. Cambridge: Cambridge University Press, 1983.

13. These quotations are from "Five Parables" in R. Rorty, J. B. Schneewind and Q. Skinner (eds.) *Philosophy in History*, (Cambridge: Cambridge University Press, 1984), pp. 118-9, where Hacking summarizes the thesis of his 1983 book.

14. *Representing and Intervening* (Cambridge: Cambridge University Press, 1983.) p. 262.

15. *Ibid.*, p. 254.

16. Following Dudley Shapere, "The Concept of Observation in Science and Philosophy", *Philosophy of Science*, vol. 49 (1982), pp. 485-525.

17. In a personal communication to the author, Hacking underlines this: "Since I'm saddled with my witticism, 'If you can spray them, then they are real', I might as well stay saddled. But do not attribute to me the thesis that if they are real, then you can spray them. I only mean to express a certain inductive scepticism—that if we never get around to intervening, then maybe what theory leads us to believe will in the end not pan out. That is a vastly weaker thesis, and one compatible with saying that we do have (imperfect) evidence for the existence of all sorts of things that we cannot use to do anything with, yet."

18. E. McMullin, "Galilean Idealization," *Studies in the History and Philosophy of Science*, vol. 16 (1985), pp. 247-273; see sec. 4

19. Gilbert Harman, "The Inference to the Best Explanation," *The Philosophical Review*, vol. 74 (1965), pp. 88-95; see p. 89.

20. *The Scientific Image, op. cit.*, p. 4.

21. *Ibid.*, p. 5.

22. *Ibid.*, p. 73.

23. *Ibid.*, p. 24.

24. *Ibid.*

25. The other source of van Fraassen's anti-realism is quantum mechanics. But it can surely be admitted that the realist thesis is more difficult to interpret and to apply in that very special domain without compromising the overall realist thrust of retroductive inference elsewhere in natural science.

26. New York: Oxford University Press, 1983, p. 56.

27. See McMullin, "Galilean Idealization", *op. cit.*

28. *The Structure of Scientific Revolutions, op. cit.*, p. 206.

29. "A Confutation of Convergent Realism," in Leplin (ed.), *op cit.*, pp. 218-249; see p. 228.

30. For a fuller criticism, see Clyde Hardin and Alexander Rosenberg, "In Defense of Convergent Realism," *Philosophy of Science*, vol. 49 (1982), pp. 604-615. See also the papers by Boyd, Leplin, and McMullin, in Leplin, *op. cit.*

WHAT WE DON'T KNOW WHEN WE DON'T KNOW WHY

SYLVAIN BROMBERGER

I. Introduction.[1]

Some twenty years ago, I gave a paper on why-questions to this Center. The occasion was—at least for me—a happy one. When I was invited to join in this twenty-fifth anniversary celebration of the Center, it struck me that a talk on the same topic would make a fitting birthday present. So this talk too will be on why-questions. But unlike the previous one, it won't be on *answers* to why-questions. It will be on a certain predicament to which why-*questions* themselves can give rise. That predicament interests me mainly because of its consequences for something I call "rational ignorance." Therefore, I will begin with a brief discussion of rational ignorance. It is also crucially related to the fact that why-questions do not have what linguists call "traces." I will therefore go on by explaining the notion of a trace and give reasons for believing that why-questions don't have any. I will turn to the predicament itself only in the last part of the paper.

II. Rational ignorance

According to received history, Eratosthenes was the first person to come close to figuring out the circumference of the earth. You remember how he went about it. He knew that, at noon on the summer solstice, the sun would be straight overhead in Syene and that no shadow would be cast there by any gnomon. He assumed (almost rightly) that Alexandria was straight north of Syene. He also knew the distance (d) between Syene and Alexandria. He erected a vertical gnomon in Syene and another in Alexandria and, at noon on the summer solstice, he measured the angle (θ) formed by the gnomon in Alexandria and the sun ray skimming its top. He had correctly reasoned that θ would be equal to the angle at the center of the earth subtending the arc between the gnomon in Alexandria and that in Syene. Then using some version of the formula

$$(1)\ C = (2\pi \div \theta) \times d$$

he calculated that the circumference of the earth was 250,000 stades, a good approximation that we can adopt for present purposes.

Eratosthenes's discovery was not an accident. He did not stumble on the answer to "What is the circumference of the earth?" the way, for example, Roentgen stumbled on radiation phenomena or Oersted on electro-magnetic ones. Erastosthenes knew that he did not know the answer and set himself to find it. In doing so, he engaged in a pivotal human activity: the pursuit of answers to pre-chosen questions. I say "pivotal" for good reasons: human cognitive growth turns on it. Children must engage in it if they are to learn the phonemes, words, phrases, and parameters of the ambient language; we must unconsciously engage in it if our vision is to find the properties of distal stimuli on the basis of proximate ones; and, of course scientific research and pooling of knowledge wouldn't be possible without it.

An activity that central to our mental, intellectual, and communal life calls for investigation. Philosophers in particular should try to clarify how knowledge and rationality can influence it. What we need to clarify can be put a little more explicitly. Imagine an ideal but intellectually ambitious rational agent—call him Rational Ignoramus—and let Q_1 be the set of all the sound[2] questions to which Rational Ignoramus does not know the answer at t_1. Being intellectually ambitious, Rational Ignoramus wants to remove questions from Q_1, that is, wants to bring it about that at some later moment t_2 he will have the answer to some member of Q_1. Under what conditions can he rationally undertake to this?[3]

Four conditions come immediately to mind.[4]

First Condition: Representations. Rational Ignoramus must be able to formulate and to understand interrogative sentences that express members of Q_1. He must therefore have access to some system in which some members of Q_1 can be represented. How else can he target any of them?

Many things about that requirement will remain obscure until we see clearly what is distinct about the systems of representations that humans can use. One thing is clear, however: some questions are beyond the reach of some systems of representation.

Consider for instance a system that admits only single-clause surface interrogative sentences of English.[5] Such a system could not represent the question now represented by

(2) There is something whose destruction calls for explanation. What is it?

since (2) is not a single-clause surface interrogative sentence. The single clause

(3) what does the destruction of call for explanation?

does express what (2) does, but is ungrammatical.

Such a system, by the way, would in all likelihood not be able to avoid ambiguities. For instance,

(4) Who did John tell what to do when?

though a single interrogative sentence, is ambiguous in that it can be construed with "when" questioning either "tell" or "do."[6]

Another thing is clear: Rational Ignoramus may be able to formulate and to understand a question and yet not be able to formulate and understand all its possible answers. Most of us understand "Who will be the President of the U.S. in 2054?" yet cannot begin to express all the possibilities.

Second Condition: Soundness. Rational Ignoramus must know[7] of some questions in Q_1 that they are sound, that is, that they have a correct answer. This does not mean that he must know the answer! In most cases he only needs to know that two sorts of presuppositions are satisfied.

First Rational Ignoramus must know of some questions in Q_1 that their *existential* presuppositions are satisfied, that is, that their referential constituents have actual references.[8] The idea is a familiar one. The question expressed by[9]

(5) What was the age of the King of France in 1945?

presupposes that there was a King of France in 1945. There was none, and the question is not sound, has no true answer. On the other hand

(6) How are neutrons and protons bound in an atomic nucleus?

presupposes that there are neutrons, and protons, and atomic nuclei. This is true, and the question is sound, must have an answer. However that answer cannot be deliberately sought by someone who does not know that there are neutrons, protons, or atomic nuclei.

Secondly, Rational Ignoramus must know of some questions in Q_1 that their *attributive* presuppositions are satisfied. So, for instance

(7) What is the height of the Empire State Building?

presupposes not only that the Empire State Building exists, but also that it has a height. This is true, and hence (7) is sound, must have a true answer. On the other hand,

(8) What is the square root of the Empire State Building?

presupposes that the Empire State Building has a square root. It does not, and hence (8) cannot have a true answer.

There are a number of ways of looking at what I have called attributive presuppositions. I will come back to the topic later, but here is a rough sketch of what is involved. (7) expresses a question built around the predicate function

(9) . . . is the height of . . .

which pairs objects and distances. The attributive presupposition of the question is satisfied if and only if the related predicate represented by the open sentence

(10) (Ex:x a distance) (x is the height of . . .)

is satisfied at the Empire State Building,[10] that is, if the Empire State Building has height, which it does.

On the other hand, (8) is built around

(11) . . . is the square root of . . .

and would be sound only if the predicate represented by the open sentence

(12) (Ex:*x* a number)(x is the square root of . . .)

were satisfied at the Empire State Building, that is, if the Empire State Building were a square, which it isn't.

Questions are built around such predicates. Many are intimately related to what used to be called determinables, that is, height ("what is the height of . . . ?"), color ("what is the color of . . . ?"), mass ("what is the mass of . . . ?"), voltage ("what is the voltage between . . . ?") and so on, a class of notions that has for too long been ignored by erotetic logicians.

Rational Ignoramus cannot know whether a given attributive presupposition is satisfied unless he has the concept of its corresponding predicate function in his intellectual kit. That fact can severely narrow his choices. To see this, imagine what things were like before the concepts of mass, electric resistance, atomic number, standard deviation, marginal utility, constituent structure, and other technical determinables were available.

He must also have some knowledge of that function's range and domain. For example, he must know that square root is defined at numbers but not at buildings or even plants, that height is defined at buildings, but not at oceans or snakes, and so on.

The grounds for that sort of knowledge, and how it can be acquired, are far from clear. Whence do we know that all physical forces have direction and magnitude, that material objects have mass, that events may have causes, that pairs of numbers have products, that sets have cardinality?

The knowledge in many cases—perhaps in all—must be grounded on the conceptual structures in terms of which we categorize, classify, taxonomize, and otherwise group things for the projection of properties and principles. Thus it is a feature of the concept of a building that a building is the sort of thing for which there must an answer to the questions "What is its height?", "Where is it located?", "Who built it?", and so on. It is a feature of our concept of a sphere that a sphere is the sort of thing for which there must be an answer to the questions "What is its volume?", "What is its radius?", "What is the area of its surface?", and so on.[11] If anything is a priori knowledge, that sort of thing surely is.

Third Condition: Awareness. Rational Ignoramus must know of some questions in Q_1 that he does not know their answer. In other words, he must be occurrently—not merely dispositionally—attending to some questions in Q_1 through some of their representations, be aware of them as sound, and know that he lacks sufficient reasons for accepting any particular answers to them.

The third condition obviously entails the first two. But it also goes beyond them. It demands that the questions have "arisen" for Rational Ignoramus, a matter over which he himself has very little control, and which his rationality can no more influence than it can influence the flow of inner and outer events of which he is aware. And it furthermore demands that he should not mistakenly believe that he knows the answers when he does not. That too is a matter over which he may have very little control and on which his rationality has less influence than some philosophers have acknowledged: insufficient reasons can be unrecognizably disguised as sufficient ones when possible answers—including perhaps the right one—fail to come to mind, or are unrepresentable, in spite of the most assiduous and appropriate intellectual efforts.

Let us call members of Q_1 that meet the first three conditions at t_1 the questions *accessible* to Rational Ignoramus at t_1.

Even under the best of circumstances, there will be members of Q_1 that are not accessible to Rational Ignoramus. This follows first from technical considerations.[12] It also follows from predicaments that Rational Ignoramus cannot avoid. Rational Ignoramus cannot possibly know of each thing, present, past, future, nearby, and far away, that it exists. Nor can he be aware of all the attributes manifested somewhere. How likely is it, for instance, that Eratosthenes could have conjured up the notion of space curvature or of valence? And it follows from the fact that the circumstances in Rational Ignoramus's life cannot raise every member of Q_1 for him, not even every member otherwise accessible to him.

Fourth Condition: Value. Since the subset of accessible questions in Q_1 is apt to be enormous, Rational Ignoramus can't try to answer them all and must make choices. If the choices are to be rational, he must have values to maximize and costs to minimize.[13] What values? The main relevant values come under four headings. I have named each after a philosopher biased in its favor.

1. Peircean value: The value that a question gets from the expected intellectual value of its answer, that is, the pleasure, or relief, or cognitive stability expected from its answer.[14]
2. Jamesian value: The value that a question gets from the material benefits expected from its answer.[15]
3. Collingwood value: The value that a question gets from the prospect that its answer will raise new questions with expected values of their own.[16]
4. Machian value: The value that a question gets from the prospect that its answer will yield answers to other questions with values of their own.[17]

All these values obviously are relativized to Rational Ignoramus's preferences and background knowledge at the time the choices have to be made.

Machian value depends on devices that turn answers to some

questions into answers to other questions, that is, things like (1) and the algorithmic procedures associated with it.[18] Let us call such devices "value adders."[19]

Let me summarize where we are by returning to Eratosthenes, our near-incarnation of Rational Ignoramus.

Before his discovery of the circumference of the earth, Eratosthenes knew that the earth existed, knew that it was spherical, knew that spheres have circumferences, knew that there was a gnomon in Alexandria, knew that there was one in Syene, knew that at noon on the summer solstice sun rays near the gnomon in Syene would cast no shadows and therefore be on a line through the center of the earth, knew that at that very time a sun ray skimming the top of the gnomon in Alexandria would form an angle with that gnomon equal to the angle at the center of the earth subtending the arc between Alexandria and Syene. The questions

(13a) What is the sun ray-gnomon angle in Alexandria?
(13b) What is the circumference of the earth?

were accessible to him. The first had little intrinsic value, but the second had enormous Peircean and possible Jamesian value. Eratosthenes knew how to use the value adder (1). That added the value of (13b) to that of (13a). Long before, others had considered

(14) What is the shape of the earth?

which had the Collingwood value of generating (13b).

So far so good. But now consider the question

(15) *Why* is the circumference of the earth 250,000 stades?

Eratosthenes clearly did not answer that question. Nor did his discovery of the answer to (13b)—even with all the rest of his knowledge—put him in a position to answer that why-question. But did it at least make that why-question accessible to him? Did it put him in a position to see that it was sound (assuming that it is sound), that it had an answer (assuming that it has an answer)? It obviously didn't.

What else did he need to know? Was he at least in a position to figure *that* out? We know that he wasn't! But why-questions must then be different, must stand in a different relation to their presuppositions. What is the difference?

Linguistic facts provide an important clue, namely that why-questions bind no traces. Let us then look at that clue.

III. TRACES AND WHY-QUESTIONS.

A: Traces.

And to begin with, what is a trace?

Think of a trace on the analogy of a variable bound to a quantifier.

(16) John hit someone.

for many purposes is analyzed as having the form

(17) (Ex) (John hit x)

Think then of

(18) Who did John hit?

as having the form

(19) (Who t) (did John hit t)

where the t in mid-sentence is the trace of "who" and is bound to it as x is bound to 'Ex' in (17). Note that the x and the t—that is, the bound variable and the bound trace—both occur in a position dictated by meaning considerations: the position in which the target of "hit" is indicated. That is the interpretive job of that position. The interpretive job of the operator's positions—that is, of the position occupied by '(Ex)' and '(Who t)'—is to indicate scope.

Traces are associated syntactically with a "movement" of the interrogative operator. In the example above, (18) and (19) derive from the deep structure[20]

(20) John hit who

—in which "who" is in a position indicating the target of "hit"—through a transformation that "raises" "who" to a sentence-initial position but that leaves a trace bound to it. The trace preserves the information that the target of "hit" is the questioned.[21] It plays therefore a semantic role.

Traces have as much reality as real words. They not only affect what an interrogative sentence means, but, like real words, they can affect pronunciation. Chomsky somewhere[22] brings out evidence of the following sort:

The interrogative sentence

(21) Who do you want to eat?

is ambiguous. It can express either a question that can take (22a) as answer, or one that can take (22b) as answer:

(22a) I want to eat the missionary.
(22b) I want the missionary to eat.

The ambiguity is easily explained: "Who" can "start out" *in situ*—that is, in a location indicating what is being questioned—as either the theme or the agent of "eat." And (21) may thus be the perceptual shape of either (23a) or (23b)

(23a) (Who *t*) (do you want to eat *t*)
(23b) (Who *t*) (do you want *t* to eat)

can only mean (23a), it cannot mean (23b). This is accounted for if the trace in (23b) stands like a real word between "want" and "to" and blocks their contraction.[23,24]

English interrogative sentences then contain mid-sentence traces in their surface structure. But they also contain traces in their *logical form*. To see this, let's look again at quantified sentences.

(25) Someone is loved by everyone.

is ambiguous: it can be interpreted either with "someone" in the scope of "everyone" or with "everyone" in the scope of "someone." We account for this by distinguishing between surface structure and logical form, and by imputing to (25) two logical forms, viz., to a first approximation

(26a) (Ey) (x) (y is loved by x)
(26b) (x) (Ey) (y is loved by x)

in which the ordering of the quantifiers indicates their relative scope. Scope ambiguities of surface structures then indicate the existence of logical forms.

But interrogative operators can also be the source of scope ambiguities. For instance,

(27) Where did everyone go?

can be construed as presupposing that everyone went to one place, or as not presupposing it. We account for this by imputing two logical forms to (27) also, namely, to a first approximation,

(28a) (Where t) (x) (x went to t)
(28b) (x) (Where t) (x went to t)

in which the ordering of the quantifier and the interrogative operator indicates their relative scope. The account obviously requires that traces occur in the logical form of interrogative sentences as well as in their surface form.[25]

English interrogative sentences then have traces in both their surface structure and their logical form. That is not true of interrogative sentences in every language. In Chinese, for instance, no traces of interrogative words occur in surface structures: the interrogative words corresponding to our "who," "where," "what," "when," and so on, remain *in situ* there; they never move to the sentence-initial position. In other words, they behave very much like quantifiers in

English,[26] which occur *in situ* in the surface structure and "move" only in the logical form. But there is evidence that Chinese interrogative operators do move in logical form. For instance, they can be the source of scope ambiguities that English interrogative operators never produce.[27] In view of these ambiguities, Chinese sentences must have logical forms distinct from their surface structure. More to the point, these logical forms, when interrogative operators are present, must contain traces. How else could they mark both the relative scope of the operators and the position questioned?[28]

B. No Traces In Why-Questions.

I said that why-questions don't have mid-sentence traces.[29] I can now be more specific: first, they do not have mid-sentence traces bound to "why" in their surface structure, and, second, they don't even have mid-sentence traces bound to "why" in their logical form. Furthermore, since translation across languages must share binding relations and relative scopes in the logical forms, it is a good bet that their expression in other languages doesn't have "why"-bound (or whatever the word for "why" in the other language) traces in the logical form either.[30]

But what reason is there to believe that "why" does not bind mid-sentence traces?

Here is one well-known argument.

Both

(29a) Who did John hit where?

and

(29b) Where did John hit who?

are acceptable sentences.

The explanation is straightforward: (29a) and (29b) both start with the deep structure

(30) John hit who where

(29a) results when "who" is raised to the sentence-initial position,[31] and (29b) results when "where" is raised to the sentence-initial position.[32] It happens that in English only one wh-operator can be raised to a sentence-initial position at surface structure. The other wh-operators, if any, must stay *in situ*. Therefore either "who" or "where" (though not both) can leave a trace and move to the sentence-initial position, and either (though not both) can end up in a mid-sentence position.

On the other hand,

(31a) Why did John hit who?

and

(31b) Who did John hit why?[33]

though at first sight related exactly like (29b) and 29a), are not both acceptable sentences. (31a) is acceptable, but (31b) is not.

The explanation is straightforward if "why" cannot occur in mid-sentence position in deep structure; in other words if there is no deep structure (32)

(32) John hit who why

but there is a deep structure (33)

(33) Why John hit who

The unacceptability of (31b) then follows from the fact that since "why" never occurs in mid-sentence, it cannot be "left" there while "who" moves to the front of the sentence. The acceptability of (31a) follows from the fact that "why" can occur in sentence-initial position in deep structure and can occupy that position in (31a) without having to "move" into it.

Since wh-words in surface structures bind traces only in positions from which they have moved, it also follows that "why" does not bind a mid-sentence trace in (31a). And from general considerations it follows that it does not bind a mid-sentence trace in any surface structure.[34,35]

The argument shows that "why" does not bind a mid-sentence trace in surface structure. I now want to turn to an argument which shows that "why" does not bind a trace in logical form. That is of greater import for Rational Ignoramus since it carries cross-linguistic consequences. The argument is also philosophically important. It is based on data that have deeply shaped recent philosophic conceptions of what why-questions are all about. In fact, the argument occurred to me while I was mulling over examples used by van Fraassen in *The Scientific Image*[36] and viewed by him as proving that all why-questions have an essential indexical aspect. To underline this connection, I will use an example taken from the discussion of why-questions in that book.

Consider

(34) Why did Adam eat the apple?

(34) can be uttered with different intonation contours. It can be uttered with a normal intonation contour or with emphatic stress on one of the words, as below:

(35a) Why did Adam eat the apple?
(35b) Why did ADAM eat the apple?
(35c) Why did Adam EAT the apple?
(35d) Why did Adam eat the APPLE?

These differences affect the question expressed in two ways.

First, different placements of emphatic stress create different implicatures. So, (35b) implicates that someone else than Adam might have eaten the apple but didn't. (35c), on the other hand, implicates that Adam might have done something else to the apple but didn't. (35d), in turn, implicates that there was something else that Adam might have eaten but didn't. (35a) has none of these implicatures.

Second, different placement of emphatic stress imposes different conditions on what counts as an answer.[37] For instance (35a) can tolerate as an answer

(36) Because God intended that to happen.

(35b) can tolerate

(37) Because he (Adam) was the one that Eve worked on.

(35c) can tolerate

(38) Because he couldn't think of anything else to do with it.

(35d) can tolerate

(39) Because it (the apple) was the only food around.

Compare this with non-why-questions.

Non-why questions, of course, can also be uttered with normal intonation contour or with emphatic stress placed on one of the words. Take "when"[38] for instance. We have

(40a) When did Adam eat the apple?
(40b) When did ADAM eat the apple?
(40c) When did Adam EAT the apple?
(40d) When did Adam eat the APPLE?

But these different placements of emphatic stress don't have the consequences they have for why-questions.

First, though emphatic stress still creates implicatures that vary with the location of the stress, these are not the ones created for why-questions. Thus (40b) implicates that others also ate the apple,[39] not that they could have but didn't. (40c) implicates that Adam also did other things with the apple, not that he could have but didn't. (40d) implicates that Adam ate also other things, not that he could have but didn't.

Second, different placements of emphatic stress do not impose different conditions on what counts as an answer. Thus (40a), (40b), (40c), (40d) all tolerate the answer:

(41) At 4 p.m. on July 7 24,000,000 B.C.

or any other answer that one of them tolerates. And they must all share the same correct answer, if there is one.

How shall we account for all this?

First note that English declarative sentences too can be produced either with normal intonation contour or with stress placed on one of the words, and that stress placed on one of the words results in a sentence logically equivalent to a so-called cleft sentence.[40] So (42a) is logically equivalent to (42b), and (43a) is logically equivalent to (43b):

(42a) ADAM ate the apple.
(42b) It was Adam who ate the apple.

(43a) Adam ate the APPLE.
(43b) It is the apple that Adam ate.

To a first approximation,[41] (42a) and (42b) share the logical form

(44) ( ){(E*x*: *x* = Adam){( ) (*x* ate the apple)}}

whereas (43a) and (43b) share

(45) ( ){(E*x*: *x* = the apple){( ) (Adam ate the apple)}}

The blanks between parentheses indicate sentence-initial positions. Those are the positions into which interrogative words, if any, are "raised." Because they are also the positions in which complementizer (words like 'that' which initiate subordinate sentences) would occur, linguists label these positions COMP (for 'complementizer').

With these analyses of declaratives in mind, we can look at our when-questions again.

Their logical forms must be similar to those of the declarative sentences, but with the interrogative operator 'when' and its trace added. The logical form of (46a), in which there is no emphatic stress, is, to a first approximation, simply (46b), that of (47a) is (47b), and that of (48a) is (48b):

(46a) When did Adam eat the apple?
(46b) (When t) (Adam ate the apple at t)

(47a) When did ADAM eat the apple?
(47b) ( ) {(Ex: x = Adam) {(when t) (x ate the apple at t)}}

(48a) When did Adam eat the APPLE?
(48b) ( ) {(Ex = the apple) {(when t) (Adam ate x at t)}}

These analyses account for the differences among the implicatures. (46b) does not have a focusing operator; (47b) and (48a) have different ones.

They also account for the common answers: all the constituents under the scope of "when" have the same interpretation since their referential expressions and variables have the same assignments.

Note that in each case, "when" binds a mid-sentence trace, for that turns out to be crucial.

There are two sentence-initial positions—two COMPs—in the logical forms of the stressed when-questions, one with the focusing operator under its scope, and one under the scope of that operator; or, to put things a little differently, one that prefixes the whole logical form, and one that prefixes the inner clause only. "When" cannot occur in the former because it binds a trace. Traces of wh-operators are anaphors, that is, they behave like reflexive pronouns and must be in the same innermost clause as their antecedents.[42] If "when" were in the outermost initial position, it would be too far from its trace, it would be outside the innermost clause containing its trace, outside the domain it must cohabit with its trace.

Now let us return to why-questions.

The logical form of (49a), in which there is no emphatic stress, is, to a first approximation, simply (49b), that of (50a) is (50b), that of (51a) is (51b):

(49a) Why did Adam eat the apple?
(49b) (Why) (Adam ate the apple)

(50a) Why did ADAM eat the apple?
(50b) (Why) {(Ex: x = Adam) {( ) (x ate the apple)}}

(51a) Why did Adam eat the APPLE?
(51b) (Why) {(Ex: x = the apple) {( ) (Adam ate x)}}

Note that "Why" occurs in each case in the initial position (COMP) of the whole sentence. Nothing prevents it from occurring there, nothing requires that it be within the innermost sentence, since, unlike "when," it binds no trace.

The difference among the implicatures, again, is accounted for by the differences among the focusing operators. (49a) does not have any; (50b) and (51b) have different ones.

The difference between these implicatures and those created in the corresponding when-questions is accounted for by the fact that the focusing operators in (50b) and (51b) are within the scope of "why," whereas in (47b) and (48b) they are outside the scope of "when." As a consequence, in the why-questions, the focusing operators serve to specify an aspect of the subject matter under question, to attach uniqueness to a presupposed item. In the when-questions, they do not serve to specify an aspect of the subject matter under question: they serve as an item in the context in which that subject matter is represented, and convey an attitude of the speaker; they attach uniqueness to an interest of that speaker.[43]

The difference among the answers called for by the different why-questions is also accounted for by the occurrence of the focusing operators within the scope of "why": since in that position they serve to specify an aspect of the subject matter under question—of the presupposition—differences among them become differences of subject matters specified, of presuppositions; and different presuppositions call for different answers.

So, everything is accounted for on the assumption that "why," unlike "when," has no mid-sentence trace in logical form.

However, one point remains. The argument hinges on the assumption that "why," unlike "when," occurs in the outermost COMP. What prevents it from occurring in the innermost COMP?

Nothing!

"Why" *can* occur in the innermost COMP. But when it does, we get a set of why-interrogatives whose implicatures and answers are like those of the when-interrogatives. To see this, picture a situation in which the speaker believes that Adam, and Eve, and God all ate the apple and asks "Why did ADAM eat the apple?" thereby indicating that he does not care about the others. Then picture a situation in which a speaker believes that Adam ate the apple, the banana, the orange, and the hamburger, and asks "Why did Adam eat the APPLE?" thereby indicating that he does not care about the rest of Adam's meal.[44]

The crux of the argument was not that emphatic "why" cannot behave like emphatic "when" (or "how," "where," and so on) but that it can behave in ways that emphatic "when" (and the others) can't.

IV. Rational ignoramus and why-questions.

The fact that why-questions have no mid-sentence traces has consequences for Rational Ignoramus.

Let us briefly return to the topic of attributive presuppositions.

Example (7) (repeated as (52))

(52) What is the height of the Empire State Building?

has the logical form

(53) (What *t*: *t* a distance) (*t* is the height of the Empire State Building)

and the attributive presupposition whose logical form is

(54) (E*x*: *x* a distance) (*x* is the height of the Empire State Building)

The connection between the attributive presupposition and the question is transparent when we look at their logical form: the only difference between (53) and (54) is their binding operator.

This means that the attributive presupposition and the question are mutually deducible by anyone who has access to these logical forms; and anyone who understands them has that access.

The transparency depends critically on the fact that the interrogative operator in (53) binds a mid-sentence trace. It is the mid-sentence trace in (53) that is the analogue of the bound variable in (54). The mutual deducibility of the question and its attributive presupposition therefore also depends on the fact that the interrogative operator binds a mid-sentence trace.

Every question whose interrogative operator binds a mid-sentence trace—that is, every question initiated in English with "when," "where," "how," "which," "what," "who," and so forth—is transparently connected in this way to its attributive presupposition.[45] Each of them has a logical form that fits the schema (55a) and a matching attributive presupposition that fits the schema (55b)

(55a) (E*x*: *x* . . .) (. . . *x* . . .)
(55b) (Wh *t*: *t* . . .) (. . . *t* . . .)

As a consequence, in every case, attributive presupposition and question are mutually deducible by anyone who understands the question.

But in every case mutual deducibility critically depends on the fact that the interrogative operator binds a mid-sentence trace, since it is the trace bound to the interrogative operator that matches the variable bound to the quantificational one.

So, as long as he limits himself to these sorts of questions, Rational Ignoramus only needs access to their logical form—that is, only needs to understand them—and his wits, to figure out their attributive presuppositions. The same is true for their existential presuppositions, since logical forms display purported reference.

In short, unassisted wits and cogitation on meaning[46] alone will tell him what he has to find out to know whether the question is sound, and thus to know whether it is accessible to him.[47]

But nothing of the sort follows for why-questions. The interrogative operator in their logical form does not bind a mid-sentence trace, and nothing stands to their logical form as (54) stands to (53).

Unassisted wits and cogitation on meaning alone, therefore, will not enable Rational Ignoramus to figure out the attributive presupposition of a why-question, even when he understands it. Or, more precisely, unassisted wits cannot lead him to that attributive presupposition in the way it can in the case of questions with bound traces. But there is no reason to believe that it can do so in any other way.

We could draw a different conclusion from the fact that why-questions do not have mid-sentence traces, namely that why-questions simply don't have attributive presuppositions at all. That is another way of looking at the fact that nothing stands to any why-question as (54) and (53) stand to (52).

But that way of looking at the facts is of little help to Rational Ignoramus. It may force me to relabel presuppositions, but it does not change the fact that why-questions have presuppositions that cannot be discovered through cogitation on meaning alone. This can be shown very quickly.

Why-questions do have presuppositions that Rational Ignoramus can discover through cogitation on meaning alone. For instance,

(56) Why is the circumference of the earth 250,000 stades?

cannot have a true answer unless the circumference of the earth is 250,000 stades. To know whether (56) is sound, Rational Ignoramus must know whether that presupposition is true. But he can obviously figure out that presupposition from his understanding of the question, that, is from its logical form.

What is true of (56) is generally true. Every why-question presupposes the proposition represented by deleting the why-operator from its logical form. Let us call that kind of presupposition the *propositional presupposition*. Whether we classify propositional presuppositions of why-questions with existential presuppositions of other questions is immaterial here.

But why-questions can have a true propositional presupposition and still not be sound. To believe otherwise is to believe that semantics alone can rule out brute facts, random events, things coming about without rhyme or reason, occurrences happening because of nothing in particular, capricious actions, ultimate truths, so's without why-so's. Semantics may be the handmaiden of rational theology, but it is not revealed theology! It cannot require that (56) have an answer.

So, to know that a why-question is sound, Rational Ignoramus must not only know that its propositional presupposition is true. He must know something else. But there is no reason to believe that cogitation on meaning alone can tell him what that something else is. (Whether we call that something else "attributive presupposition" is immaterial again.)

How can that be?

A full account would require a detailed discussion of the conditions that define true answers to why-questions. I am not prepared to do that here or now. However, many a blind person—myself included—has already reported on the shape of that elephant, and its general outline is beginning to emerge.

A why-question can have a true answer only if its propositional presupposition expresses a fact. But that fact in turn must be construable as the answer to a question which is linkable to other questions by means of value adders, and not just any value adder, but a value adder that occupies a particular position in the hierarchy of value adders.[48] However, Rational Ignoramus cannot always know in advance whether a given fact can be construed that way, or how it is to be construed that way. The right value adders, predicates, determinables, and attributes must first be conceived and checked. The logical form of the why-question is no guide to these. It does not even constrain them. It has no direct conceptual relation to them. In fact, a why-question may be expressible long before the relevant value adders and attributes. "Having mass," "having outer electrons," "having a logical form," "being a combination of distinctive features" and so forth, and the value adders that link them to other determinables were conceived and received expression long *after* why-questions that called for them had been raised. The logical form of these why-questions was of no help to those who raised them. It

Where does that leave Rational Ignoramus?

In the very non-plussing predicament of knowing that he does not know the answer to some why-questions, yet not knowing what he must find out in order to know whether there is anything to know at all![49]

What is Rational Ignoramus to do under the burden of such ignorance? My guess is that the rational thing for him to do is to forget about the why-question and to turn to other questions instead, remembering that answers to why-questions usually emerge from work on questions with more reliable credentials. That same guess probably looms behind the distrust of why-questions often felt by experimentalists, and behind the eagerness of some philosophers to reconstrue why-questions as non-why-questions in disguise, for example, as "What is the cause of . . . ?", "From what laws and antecedent conditions does it follow that . . . ?", "What mechanisms are responsible for . . . ?", "For what reason did . . . ?".

But even if my guess is right, it should not be taken as denigrating answers to why-questions. Those answers include some of our most beautiful intellectual prizes. Answers to why-questions reveal the surprising in the familiar and the familiar in the surprising, they unveil what is otherwise hidden, they link what seems otherwise unconnected; they generate new questions; they come with new value adders. The credentials of why-questions may be unclear, but their Peircean, Jamesian, Collingwood, and Machian value is beyond doubt. That is probably why distrust of why-questions has never influenced theory construction for very long, and why philosophical reconstruals have never gained wide acceptance.

NOTES

1. I thank Ken Hale, Morris Halle, Jim Higginbotham, Richard Larson, Guiseppe Longobardi, and David Nash for much of the information in and behind this paper, and for other forms of help, Tom Kuhn for comments on a previous draft and for many useful discussions around these topics, and Nancy Bromberger for editorial and orthographic advice. I owe much to Jaakko Hintikka's work on topics related to what I call "rational ignorance" and to our conversations about the subject.

2. By a sound question, I simply mean a question that has a correct, true answer.

3. By defining Q_1 as made up of only sound questions we leave out of consideration, for the present, problems caused by the fact that Rational Ignoramus must also cope with questions he mistakenly believes to be sound.

4. These are not the only ones, but they are the only ones that matter for present purposes.

5. English is *not* such a system.

6. Ambiguity of interrogative sentences raises a vexing problem. To say that an interrogative sentence is ambiguous is to say that it can be used to express more than one question. If there are ambiguous interrogative sentences—as there obviously are—this means in effect that we cannot identify a question by just any sentence that expresses it. How then shall we identify questions? In the case of declarative sentences helpful notions of ambiguity are available: for example, that a declarative sentence is ambiguous if and only if it can be evaluated through more than one set of truth-conditions. This suggests that what a declarative sentence expresses may turn out to be identifiable by a set of truth conditions. Working out this notion is fraught with difficulties. But it gives us some sort of grasp. Nothing of the sort is available for interrogative sentences: interrogative sentences do not have truth conditions. Identifying them through the truth conditions of the declarative sentences that express their possible answers or their presuppositions doesn't work either. In a later section I will presume that these sentences have unambiguous logical forms.

7. "Know" here and elsewhere may seem too strong. "Believe" or "believe on good grounds" may seem better. These subtleties are beyond the accuracy of this discussion.

8. This way of putting the point implies that questions can have referring constituents, that is, are sentence-like objects. I happen to think that this is true, but the condition can, in any case, be restated so as to mention only constituents of the sentences used by Rational Ignoramus to express questions. The restatement is joylessly convoluted and forces issues about ontological and conceptual relativity to the surface. I will bypass it.

9. From here on, I will drop this awkward "the question expressed by" and let context indicate whether the example is to be taken as a question expressed by the sentence or the sentence itself.

10. The use of a (restricted) existential quantifier in (10) may suggest that we are simply dealing with another existential presupposition. Note that (7) can be expressed with "how" rather than "what" ("How high is the Empire State Building") to mark the difference between existential and attributive presuppositions. Which is which in specific cases cannot be settled without settling general and specific issues of ontology and ideology.

11. But there can be conceptual changes. It is no longer a feature of the concept of physical object that a physical object is the sort of thing for which there must be an answer to the questions "What is its mass?", "Where was it located when?", "What was its velocity relative to what frame of reference when?", and so on.

12. For instance, there are as at least as many questions as there are real numbers—that is, nondenumerably many—since there are questions of the form

(57a) What is its square?
(57b) What is the prime number closest to it?
(57c) What is the difference between it and the average population of New York in 1924?

for each real number. But the total number of questions expressible in Rational Ignoramus's language can be at most denumerably infinite, since each must be expressed in a finite structure constructed out of a finite—or at best denumerably infinite—vocabulary. Thus, after taking into account all the expressible questions to which Rational Ignoramus knows the answer (a denumerable set), we will still be left with nondenumerably many sound questions.

13. I omit discussion of costs and of certain other relevant considerations such as likelihood of success in the pursuit of an answer.

14. Elsewhere I have called this "gosh" value.

15. Elsewhere I have called this "cash" value.

16. Elsewhere I have called this "golly" value.

17. Elsewhere I have called this "added" value. Naming it after Mach is a little misleading. I do so because he valued laws and theories for the "intellectual economy" they made possible. That comes close to valuing them as means for yielding what I call Machian value to questions.

18. (1) gave added value to both "What is the value of θ?" and to "What is the value of d?" The added value given to each of these depended, of course, on the prospects of getting the answer to the other.

19. Intellectual value adders differ along such dimensions as precision, reliability, accuracy. But they also differ in power. Consider for instance

(58a) All elephants have warm blood.
(58b) Is Jumbo an elephant?
(58c) Is Jumbo warm-blooded?

The value adder (58a) adds the value of (58c) to the value of (58b) only if the answer to the latter is "yes." Otherwise it adds no value to (58b). It is a conditional value adder. Contrast that with

(59a) The distance between Memorial Hall and Kresge is twice the distance between Central Square and the middle of Harvard Bridge.

(59b) What is the distance between Memorial Hall and Kresge?

(59c) What is the distance between Central Square and the middle of Harvard Bridge?

(59a) adds the value of (59c) to the value of (59b) no matter what the answer to the latter. It adds value unconditionally. And now consider

(60a) $T = 2\pi \sqrt{1/g}$

(60b) What is the length of the pendulum built by Foucault?

(60c) What is the period of the pendulum built by Foucault?

(60a) adds unconditionally the value of (60c) to the value of (60b). But, more important, it adds to any question of the form

(61) What is the length of . . . ?

about some pendulum, the value of the corresponding question of the form

(62) What is the period of . . . ?

about the same pendulum.

We have here the beginnings of a hierarchy that extends to the farthest levels of theory.

20. The expressions "deep structure," "surface structure," and "logical form" are shunned nowadays because they call up misleading associations. Linguists prefer to use "D-structure," "S-structure," "L-form." I will use the old terminology in the belief that some mnemonic association is better than none.

21. Linguists use the term "raise" because the transformation moves "who" from the end of a lower branch to that of a higher one in a tree-like representation of the sentence's structure. Other transformations—of no interest here—are also required to get (18) and (19): one that inserts "did," for example. All this talk of transformation is not to be taken literally: it is a picturesque way of talking about relationships among various levels of analysis.

22. Noam Chomsky, "On Wh-movement," in Culicover, P. W. et al. (eds.), *Formal Syntax* (Academic Press, 1977)

23. This argument must be handled with extreme caution: traces are "visible" and block contractions only under specific conditions. Many

traces do not block contraction. But for present purposes, it is sufficient that some do.

24. The data are from a dialect of English that does not have "whom." Speakers of dialects that have "whom" may find "What do you want to smell?", "I want the flowers to smell," "I want to smell the flowers," "What do you wanna smell?" more convincing. But since their dialect probably does not tolerate "wanna" in any case, they should look at the data as taken from a dialect that happens not to be theirs. That does not make it less real.

25. There are many other grounds for holding that sentences have logical forms as well as surface and deep structures.

26. Or like all but one wh-operator in English multiple questions. For example, "Who went where, when, how." See (29a) below.

27. The sequence "John know his mother see who?" matches the sequence of words in a Chinese sentence that can be interpreted as either "Who does John know that his mother saw" or "John knows who his mother saw."

28. Everything has to be qualified! We have known ever since Schönfinkel that sometimes combinators properly arranged can perform the two corresponding jobs for quantifiers without binding anything. The same is probably true of traces. Offhand, I don't know how to handle that complication and save the rhetorical point of my question. But the thesis that why-questions have no mid-sentence traces should translate into an equivalent thesis in a scheme using combinators, if it turns out that the logical forms of sentences in a natural language are representable with combinators.

29. I use "mid-sentence" out of technical caution. As far as I know, "why" may bind traces in sentence-initial positions, that is, in what will be called COMP below.

30. Another qualification! "How come" questions do have a trace bound to "how" after "come." Doesn't that pose a problem if—as seems plausible—"(how *t*) (come *t* that *p*)" turns out to be a synonym of "why *p*," or if some expression much like it turns out to be the translate of "why" in some other language? Not really. The argument below hinges on the absence of a trace bound to "why"—or to any of its translates—inside of *p*—or any of *p*'s translates. The innards of *p* are what matters. I thank Ken Hale for helping me with this point.

31. When the trace is made visible, (29a) is "(Who *t*) (did John hit *t* where?)."

32. When the trace is made visible (29b) is "(Where *t*) (did John hit who (at) *t*?)."

33. Not to be confused with "Who did John hit and why?".

34. The account also implies that "why" cannot carry the sort of thematic role that arguments of verbs or prepositions carry.

35. This argument, being an inductions to a (best?) explanation, needs strengthening from further evidence. Such evidence exists. For instance, Giuseppe Longobardi, in a conversation, has shown me a beautiful argument based on the observation that whereas both members of a pair like

(63a) Why do you think John left?
(63b) Why do you think that John left?

—construed as about the reasons for John's leaving, not for thinking that he left—are acceptable, only (64a) of the pair below is acceptable:

(64a) Who do you think went to London?
(64b) Who do you think that went to London?

Yet, at first sight, the difference between the members of both pairs is the same, namely the absence or presence of 'that'.

36. Bas van Fraassen, *The Scientific Image* (Oxford: University Press, 1980) Chapter 5, Section 2.

37. The fact that stress affects what counts as an answer has wrongly been taken to show that the construal of why-interrogative sentences like (34)—even if we abstract from the presence of explicit indexicals—is more crucially dependent on context of utterance than that of other sentences. It shows nothing of the sort. What it does show is that accentuation is one of the individuating features of sentences. It has also been taken to show that (35b), (35c) and (35d) express different ways of construing (35a). They don't. Their readings are not among the readings of (35a) even if—as some believe—the answers they tolerate are among the answers that (35a) tolerates. (The converse is not true, as lovers of poor jokes well know.) (35b) is no closer to expressing a way of construing (35a) than the cleft interrogative sentence "Why was it Adam who ate the apple"—with which it shares logical form—is. (35a), (35b), (35c), (35d) are different sentences each with its own construal. (35b), (35c), (35d)—though not (35a)—happen to share logical form with clefted interrogative sentences, and each therefore expresses the same question as its corresponding clefted interrogative sentence. But interrogatives with different logical forms express different questions, regardless of the similarities in their surface structure. The point may be obvious but it is overlooked by theories that take why-questions with emphatic stress (or clefted) as paradigmatic cases of why-questions, and that do so on the mistaken presumption that every why-question can be expressed with emphatic stress on one of the words, or by a cleft interrogative. Such theories start off on the wrong foot in any case, since their inductions rest on biased samples with idiosyncratic logical forms.

38. "How" or "where," and so on, would do just as well.

39. Admittedly *that* implicature is rather difficult to swallow.

40. This is not the only instance in which departure from normal intonation contour changes logical form. Declarative sentences produced with an interrogative intonation contour have the logical form of an interrogative sentence.

41. I say "to a first approximation" because many details are left out and because the "(Ex: $x = N$)" representation of the focusing operator is crude.

42. That this is so for reflexives is illustrated by the contrast between (65a) and (65b):

(65a) John helped himself
(65b) John asked Mary to help himself

In (65b), unlike in (65a), "himself" is not in the same innermost clause as "John." The full story requires further elaboration of what counts as a clause here.

43. They function like parentheticals.

44. People with sharp ears will notice a slight change in the tonality of the stress. That fact too constitutes confirming evidence: we know that logical form and phonology interact.

45. Pesetsky and others have recently suggested that some of these English operators do not move into a COMP in logical form but remain *in situ* and that the same holds in other languages. My main point should not be affected if they turn out to be right, though I would then have to complicate the presentation. My point rests on the presumption that "why" is different from the others in that it never occurs *in situ* in mid-sentence—and hence never has a trace there—whereas the others either occur *in situ* or have a trace there.

46. "Meaning" makes some people cringe. But it shouldn't in this context where it carries very minimal commitments.

47. Cogitation on meaning will often also tell him *whether* an attributive presupposition is satisfied: as I mentioned a while back, that sort of knowledge often springs directly from conceptual structure, is *a priori* if anything is.

48. See note (19) above.

49. Why-questions are not unique in this respect: "What-is *X*?" questions, under certain circumstances, will raise the same predicament, but in a different way. There must be others. The topic deserves investigation. This predicament, by the way, is not to be confused with what I once called the *p*-predicament, viz. the predicament in which one is with regard to a question *q* when one knows that the presuppositions of *q* are true but every

possible answer to q one can excogitate is incompatible with other things one knows. Why-questions can put one in either predicament. Questions whose presuppositions entail that at least one possible answer is true can only put one in a p-predicament.

MENTAL REPRESENTATION: AN INTRODUCTION

J. A. FODOR

THIS paper is meant to introduce you to a new theory about how the mind works. Generally speaking, theories have certain logical properties in common with headaches. For example, every theory, like every headache, tends to be somebody's theory or other. This is useful, in the case of headaches, because once you know whose headache a headache is, you know who to give the aspirin to. And it's useful, in the case of theories, because once you know whose theory a theory is, you know who to go to to find out what it is that the theory claims. And who to blame in case something goes wrong.

The theory I want to tell you about, however, is an exception to this rule. It has, so far as I can tell, no definite provenance, having developed as a sort of epiphenomenon of work in a number of different fields; most notably linguistics, computer science, logic, psychology and philosophy. In consequence, there isn't any place you can go for the official, authorized version. In fact, it's only in the last five or ten years that the theory has come to have an official, authorized name. It is now widely agreed that, whatever exactly the theory *is*, the right thing to call it is: COGNITIVE SCIENCE. So, what I shall be giving you in this lecture is an account, not official and not authorized, but nevertheless accurate as far as I can manage, of what cognitive science is.

I shall be emphasizing repeatedly in this talk respects in which cognitive science is a recidivist account of the mind. In my view, much of cognitive science is the rediscovery of doctrines that were familiar—in fact, commonplace—in the tradition of classical epistemology that ran, say, from Descartes to Kant with intermediate stops at Locke, Berkeley and Hume. Indeed, it is a curiosity of the cognitive science movement that it has significantly more in common with the psychological theorizing of the 17th, 18th and 19th centuries than it does with the psychological theorizing of the first half of the 20th century; so that many of the characteristic doctrines of the movement would have been more agreeable to, for example, John Stuart Mill or William James than to Watson, Hull, Thorndike or, of course Skinner.

But while I think it's right to view cognitive science as carrying forward an old philosophical tradition, there is one respect in which its doctrines are a radical departure from much of what preceded it in the philosophy of mind. *Almost* all the great philosophers who wrote about the mind had a prejudice in favor of *conscious* mental states; indeed, a substantial number of them—notably those in the Cartesian tradition—took consciousness to be an *essential* property of the mental; so that, on their view, the notion of an unconscious mental state is simply self-contradictory. It's not entirely clear why the Classical tradition clung so tenaciously to the consciousness of the mental, but a reasonable guess might be that it connects with another doctrine that many of the major figures shared: viz. that a theory of the mind should do double duty as part of a foundationalist epistemology. For example, theories of perceptual integration were supposed to illuminate the *justification* of perceptual beliefs. On that view, stages of mental processing correspond closely to steps in justificatory arguments; and the worry was that if the former were unconscious the latter would be unavailable for citation. We would thus be without reasons to give for the the perceptual beliefs that we hold.

Anyhow, cognitive scientists tend to view foundationalist epistemology with suspicion and consciousness with loathing. The reason for the latter is simply that, so far at least, the tenets of the theory fail to illuminate the nature of consciousness; indeed, they don't *even* explain why there *is* such a thing; why any mental states *are* conscious. From the cognitive scientist's point of view, consciousness is an embarrassment; a pathological condition of the mental states that he studies.

You may wonder how anybody could hope to get away with a doctrine that perverse; and why anybody should be tempted to try to. The answer is, I suppose, that while there are some mental states that it's hard to imagine as other than states of consciousness (pains, itches and afterimages are plausible candidates) there are others, the so-called "propositional attitudes," where the connection between the mental and the conscious seems less close. (For example, beliefs and desires, unlike itches and after images, can and often do lead an entirely dispositional existence; and what is entirely dispositional is presumably ipso facto not conscious.) In point of historical fact, it is the investigation of mental states of this second sort to which serious attention in cognitive science has been exclusively directed. First Freud, and after him a legion of others, made clear how much the postulation of unconscious propositional attitudes might buy us: the possibility, to put it briefly, of extending, the explanatory purview of commonsense belief/desire psychology[1] far beyond what commonsense had ever imagined. Given the postulation of unconscious propositional attitudes, a class of phenomena ranging from verbal slips to perceptual constancies promises to assimilate to familiar explanatory mechanisms. This, surely, is cheap at the price; an offer that we can't refuse.

The picture, then, is that there is one kind of mental state, including pains, itches and afterimages, of which consciousness is an *essential* property; there *could not be* (or so the story goes) unconscious states of this sort. Philosophers sometimes call these "qualitative" states. This is, to repeat, the kind of mental state that cognitive science is *not* about. The kind of mental state that cognitive science *is* about is exemplified by beliefs and desires. The essential, defining property of states of this latter sort is that they have *intentional content*. Fundamentally, cognitive science is a theory about the intentional mental states.

What's intentional content, then? That turns out to be a little hard to say; for present purposes, let me just try to convey the basic intuitions. Intentional mental states have in common the property of being—or, anyhow, of purporting to be—*about* things. So, for example, if I believe that the moon is pretty far away, then there is something one can say be way of answering the question "What is Fodor's belief about?" Answer: "It's about the moon," (or maybe "it's about how far away the moon is," or maybe "it's about the moon's being pretty far away." Any of these will do for the purposes at hand.) Similarly, suppose I have a certain desire; viz. a desire for a unicorn of my very own. Then there is something that my desire is- or purports to be- about; viz. a unicorn or, if you like, a unicorn of my very own.

Philosophers since the Middle Ages have amused themselves by pointing out logical peculiarities that infest the relation between an intentional *state* (like a belief or a desire) and its intentional *object* (viz. the thing that it is or purports to be about.) For example, consider the difference between wanting a unicorn and riding a unicorn. The former, but not the latter, is a condition that can be achieved in face of the nonexistence of unicorns. Hence, if Fodor rides a unicorn there must be an answer to the question: "Which unicorn does Fodor ride?"; but though Fodor wants a unicorn, there need be—perhaps there can be—no answer to the question: "Which unicorn does Fodor want?" A mental state can thus be "about" an *x* even if there aren't any *x*'s. Indeed, a mental state can be about an *x* even if there *couldn't be* any *x*'s. It is possible, though arguably pointless, to want a square circle.

So, there are "qualitative" mental states and there are "intentional" mental states, the former characteristically endowed with awareness and the latter with aboutness. Traditionally, philosophical interest in the nature of intentional states has had three motivations. First, there is the desire to figure out what aboutness is about, why it displays the logical peculiarities it does, and how mental states (or, for that matter, anything else) could have it. Second, there is a specifically epistemological motivation: knowledge is itself an intentional state since knowledge is or purports to be about the world. So if you're interested in the nature of knowledge, you are willy nilly interested in the problem of intentionality. And finally—as previously intimated—the intentional mental states seem to be closely implicated in the causation of behavior; specifically, of intelligent, "higher cognitive" behavior. Theorizing about intelligent behavior is, of course, the cognitive psychologist's proprietary beat so we arrive at the first of several interdisciplinary connections that we'll come upon in the course of this survey. The contemporary focus on intentional mental states has lead to a renewal of connubial relations between psychology and the philosophy of mind after decades in which practicioners of each were extremely eager to dissociate themselves from practicioners of the other. (Tweedledeedum and Tweedledeedee were, for the longest while, furious with one another over the question whether the mind should—or can—be studied *"empirically."* It turns out, however, that nobody knows precisely what this question means; which makes it seem a lot less pressing how one answers it.)

Cognitive science shares with mere commonsense the idea that you and I (and, presumably, many other of the higher organisms) typically act out of our beliefs and desires. That is, cognitive science and commonsense endorse explanations of what people do which refer, perhaps ineliminably, to the agent's intentional mental states. Some of these explanations get pretty fancy, but for our present purposes the simplest sorts will do. Here, for example, is a piece of behavior I from time to time indulge in is: I put a stamp on an envelope in which I place a letter that I have written. What is the presumptive explanation of this behavior? Well, however, the details may go, it is surely plausible that some of my beliefs and desires play a causal role. Thus, I have beliefs about stamps, about the post-office, about zip codes and addresses, and so forth, from which it follows that there is some modest chance that a letter dispatched via the postal service will actually reach its addressee; and, moreover, that that chance is measureably increased if the letter is properly stamped. Nested in these beliefs are complexes of convictions I hold about the conventions for addressing letters and the conventions for writing English, about my geographical situation and my correspondent's . . . etc; so the story actually gets quite complicated if you try to spell it out.

So much for my beliefs. It is also part of the commonsense story about putting stamps on envelopes that the stamper has certain preferences: that he prefers a world in which the letter arrives to one in which it doesn't, and that he is willing to pay at least twenty-two cents (current rates) to live in the former of these worlds rather than the latter.

Why on Earth, you may reasonably wonder, am I telling you these platitudes. Well, for one thing, rehearsing platitudes is a fair part of how philosophers make their living. And, for another, it's true of *these* platitudes that they have, at one time or other, been denied by a number of important people. I remind you, for example, that American psychology was explicity behaviorist for some fifty years and so was committed to the view that no scientific theory of the mind could accept that behavior has mental causes. What American psychology meant when it said this was that such explanations as the one that we just walked through *could not be literally true*. Indeed, many—perhaps most psychologists and philosophers *still* hold this (though not, perhaps, on explicitly behaviorist grounds). By contrast, the cognitive science position is committed to taking this sort of etiology as both true and literal. Beliefs and desires have causal roles in the production of behavior and a psychological theory of behavior is primarily an account of these causal roles. This commonsensical acceptance of mental causation is an example of the theoretical conservatism which, as I remarked at the outset, is largely characteristic of the cognitive science movement.

The second reason for stressing the view of behavior as caused by mental states is that it allows us to see the issues about intentionality in a somewhat different context. We remarked a while back that some mental states (notably beliefs and desires) are, or purport to be, about things. We then remarked that these same mental states appear to play an important causal role in the production of behavior. But now it turns out that these two considerations are connected by the following principle: *the causal role that an intentional state plays vis a vis behavior depends upon the intentional content of that state*. Or, to put it less pedantically, what having a belief or desire causes you to do is not independent of what it is a belief or desire about. Consider the following example.

According to the well-known anecdote, somebody once asked Willie Sutton why he robbed banks. "Because that's where they keep the money," Sutton is supposed to have replied. It seems to me very likely that this was the right explanation of Sutton's deplorable behavorial proclivities; cognitive science hasn't gotten much beyond it. The point for present emphasis is that the causal role that Sutton's belief about banks played in engendering his behavior was intimately determined by its content. If, for example, Sutton had said that he robbed banks because he believed that they kept carrots there, or because he believed that two is a prime number, or because he believed that Scott was the author of Waverly, then, in all likelihood, his explanation of his behavior *wouldn't* have been right.

We know this because we know that beliefs have their causal roles in virtue of their contents and, first blush at least, it's pretty easy to see how beliefs about money could have a role in the etiology of stick-up behavior and pretty hard to see how beliefs about carrots or prime numbers could.

So now we have two themes and a connection between them. Beliefs and desires have intentional contents and they have causal roles; and the character of their casual roles is somehow closely connected with the character of their intentional contents. This seems to be a rich enough web of phenomena to demand a theory; how, one wants to know, could mental states be such that all of this is true of them? It is *precisely* this question that cognitive science purports to answer.

At the heart of the proposed solution lies the following observation: beliefs and desires aren't the *only* sorts of things that are simultaneously possessed of causal intentional properties. There's one other sort of thing (and, so far as I know, *only* one) in which both sorts of properties inhere: viz. *symbols*. Suppose, for example, that I write on the board a certain sentence: the cat is on the mat. Then consider the properties of the inscription that I produce by doing so. On the one hand, it's a physical object, a measurable weight of chalk dust distributed in a geometrically characterizable configuration. *Qua* physical object it has an indefinite complex of (largely dispositional) causal properties. For example, it exerts a certain gravitational force upon the moon, it reflects light in various complicated ways, it offers an in principle detectable resistance to the operation of blackboard erasers, and so forth. But, on the other hand, this sort of physical object is special in that it also has a bundle of intentional properties, much as we have seen that beliefs and desires do. So, in particular, it looks to be about something (viz. the cat, or where the cat is, or the fact that the cat is on the mat); and it has in common with the belief that the cat is on the mat that it is true in some conditions and false in others; in particular, the inscription is true if and only if the belief is; and both are true if and only if the contextually relevant cat is on the contextually relevant mat.[2]

This parallelism between symbols and intentional mental states is simply too striking to treat as accidental; it cries out for some sort of theoretical *reduction*. There are, patently, two possible strategies.

Either show how the intentional properties of mental states derive from the semantic properties of symbols; or show how the semantic properties of symbols derive from the intentional properties of mental states. So we arrive at our second locus of interdisciplinary connection. Just as reflection upon the causal role of intentional states discovers a nexus between cognitive psychology and the philosophy of mind, so reflection upon the similarities between mental states and symbols discovers a nexus between theories about the mind and semantic theories. Quite a lot of philosophical controversy during the last three or four decades can be viewed as a disagreement over how to understand this latter relation, and we're about to see that cognitive science is committed to one of the options.

The standard treatment of the mind/symbol relation in modern philosophy has been to attempt to derive the intentional properties of the former from the semantic properties of the latter; in particular, to attempt to derive the intentional properties of mental states from the semantic properties of *linguistic* symbols like English sentences. Thus, in the crudest sort of formulation, you might try identifying the belief that there is a cat on the mat with a disposition to *say* that there is a cat on the mat, where saying that there is a cat on the mat is in turn analyzed as involving the production of an utterance of some symbol which *means* that there is a cat on the mat (eg. the production of an utterance of the sentence "there's a cat on the mat"). The claim is that it's this "conceptual" connection between beliefs and utterance tokenings that bestows intentional content upon the former. I won't discuss the motivation for this treatment except to say that it is historically connected with a version of the behaviorist program; specifically with the so called "logical" behaviorism which sought not the *elimination* of mentalistic concepts from psychological theory, but rather their reduction to concepts which subtend only behavioral states and events.

Anyhow, the proposal seems to be hopeless. To mention just two of the standard objections: if the properties of mental states are taken to derive from dispositions to utter linguistic symbols, then we are going to have a terrible time holding onto the idea that *any* infrahuman organisms can think. The issue here is not just politeness to animals. It is natural—especially if you cherish any sort of Darwinism in the explanation of behavior—to hope that there will be some substantial continuity between animal psychology and human psychology. But it's hard to see how you could get this unless some form of belief/desire psychology applies to infrahuman creatures and hence to infralinguistic ones. Second, human beings *learn* the languages that they eventually come to speak; but learning itself seems to be a mental process—indeed, a higher cognitive process—and hence to involve thought. (We accept this without hesitation in the case of a person learning a *second* language, so it seems—to put it mildly—a little arbitrary to reject it in the case of a person learning a first one.) It is therefore hard to see how learning could take place in prelinguistic organisms if prelinguistic thinking is ruled out as, of course, it would be on the view that thoughts are anything like dispositions to utter sentences. Yet babies *are* prelinguistic organisms and the *do* learn to talk.

This objection isn't, of course, conclusive. Instead of rejecting a conceptual dependence of thinking upon talking, you could try rejecting the idea that learning involves thinking. This is, in point of historical fact, the line that many philosophers have taken. The doctrine that thinking presupposes linguistic competence is thus supported by a behaviorist (typically a learning-theoretic) account of language acquistion: the view, as Wittgenstein puts it, that language acquisition is largely a matter of "training." I'm suggesting that the relation between these views is intrinsic. You can't coherently opt for the reduction of thought to language unless you are prepared to accept some sort of behaviorist account of language acquisition. The disrepute into which the latter doctrine has now fallen is thus a reason for scrapping the former.

The alternative direction of reduction would be to derive the semantic properties of language from the intentional properties of thought. The idea—which is, in my view, highly plausible so far as it goes—is that the function of language is to *express* thought. So, for example, the sentence "the cat is on the mat" means what it does in virtue of the fact that it is the form of words that English speakers use to express a belief with a certain intentional content; viz. the belief that the cat is on the mat. This account shares a reductionist impulse with the neobehaviorist story but, to repeat, the reduction goes the other way 'round: the intentional/mental properties are basic and the semantic/linguistic ones are derived. In most current versions of this doctrine, beliefs, desires, etc. are viewed as monadic functional states; that is, as nonrelational states that are definable by reference to their typical causes and effects. It is these monadic functional states that are taken to possess, as one might say, aboutness in the first instance. There is, in particular, no more fundamental vehicle of aboutness from which the intentionality of mental states is itself supposed to be derived.

I've just been describing the standard modern construal of the thought/language problem: to specify and motivate a reductive account of the relation between the two; to show either that language is what it is because it expresses thought, or that thought is what it is because it is expressed by language. However, in the philosophical tradition prior to this century, one finds a rather different conception of how the parallelism between thoughts and symbols might be understood; a venerable third option that is just now coming back into fashion. The idea, to put it in a nutshell, is that it might be possible to pull off a *double* reduction: First derive the semantic properities of linguistic symbols from the intentional properties of mental states; then postulate a population of *mental symbols*—mental representations, as one now says—and derive the intentional properties of beliefs and desires from these. Practically every major Twentieth Century philosopher, at least in the Anglo-American tradition, has taken it for granted that such a double reduction is *out of the question*. But it increasingly appears that they were all wrong.

First, let me make clear what the proposal is. We start by observing a certain parallelism between beliefs and desires, on the one hand, and linguistic symbols on the other: both exhibit intentional and causal properties. This similarity suggests a reductive treatment; specifically a treatment in which the intentional properties of linguistic symbols are derived from the intentional properties of the thoughts that the symbols are used to express. At this stage, the intentional properties of mental states are viewed as fundamental, and the intentional properties of linguistic symbols are viewed as derived. But now it's suggested that the reduction can by carried a stage further. Suppose we think of beliefs, desires and the other intentional states (not as *monadic* but) as *relations that organisms bear to internal mental representations*. These mental representations are, by assumption, symbols; and, *qua* symbols, they have semantic properties. Since mental symbols have semantic properties, and since intentional mental states are construed as relations to mental symbols, we can think of beliefs and desires as inheriting their semantic properties from those of mental representations, just as we have been thinking of linguistic symbols (eg. English sentences) as inheriting *their* semantic properties from the intentional properties of the beliefs that they are used to express. On this view, the semantic properties of mental representations are at the very bottom of the pile. It is mental representations that have aboutness *in the first instance;* everything else that is intentional inherits its intentionality, directly or otherwise, from the intentionality of mental representations.

It may be that some of you now detect just a soupcon of infinite regress. Having gone this far, you might say, why stop? Why not posit an infinity of alternating strata of beliefs and desires underlying mental symbols and mental symbols underlying beliefs and desires? And if, indeed, there is no reason to stop, that would itself be a conclusive reason for never having started.

What's required, of course, is independent motivation for the second phase of the double reduction (i.e. for the postulation of mental symbols; I take it that the first phase is sufficiently motivated by the desire to treat the similarity of mental states and linguistic symbols as not accidental). The major motivation derives from a part of the theory that I haven't yet gotten to: the treatment of mental *processes* (processes like, for example thinking; as contrasted with *states* like, for example, belief). I'll return to make a point of this later in the discussion. For the moment, let me propose two problems that suggest subsidiary arguments for taking intentional mental states to be relations to mental symbols from which their "aboutness" is inherited, as opposed to treating them as monadic states whose intentionality is underived.

1. *The problem of "common contents."* Many intentional mental states share their contents; just as one may believe that Henry James is a better novelist than Trollope, so too one may doubt, fear, dread, remember or hope that this is so. What are we to make of this? What, in particular, is the common element in this heterogeneous collection of mental states? The problem is serious if you view mental states as merely monadic; viz as functional but not relational. For

then, given the sorts of things that functional states are, if there is something in common between, say, the hope and the dread that James is better than Trollope, this common element must presumably be some (possibly higher order) property of the causal roles of those states. But what could this common element be? What could it be (to vary the example) that both a preference and a loathing for eating chocolate ice cream cause, such that they are both about eating chocolate ice cream in virtue of their both causing that?

Whereas, consider the charm of the alternative view; beliefs, desires and the rest are relations between organisms and mental symbols, which, to put it very roughly, express the contents of the mental states. So that what it is to believe James to be better than Trollope is to be in a certain, as it were aye-saying, relation to a mental symbol which means that James is better than Trollope; and what it is to doubt that James is better than Trollope is to be in a different, as it were nay-saying, relation to that same symbol. Similarly, mutatis, mutandis, for chocolate ice cream.

2. *The problem of constituency.* Beliefs appear to have constituent structure; for example, the belief that it's raining and the streets are wet appears to be in some quite intimate way related to the belief that it's raining and the belief that the streets are wet. Problem: what is the nature of this relation?

The obvious solution is that the constituency relation for beliefs is literally a part/whole relation; on that account, the belief that it's raining is *part* of the belief that it's raining and the streets are wet. And so is the belief that the streets are wet.

What makes the constituency problem interesting, however, is that this solution doesn't work. Notice that the same sort of constitueny relation that holds for *con*junctive beliefs must also be defined for *dis*junctive beliefs like (eg.) the belief that it's raining *or* the streets are wet. And it is clear that neither the belief that it's raining nor the belief that the streets are wet can be parts of the disjunctive belief since it is possible to have the latter without having either of the former. To put it crudely, there seems to be something *P*-ish and something *Q*-ish about the belief that *P* or *Q*; and it's a puzzle what these *P*-ish and *Q*-ish things might be since they aren't either the belief that *P* or the belief that *Q*.

It throws light on this problem that there is another one just like it about speech acts. In particular: What are the constituents of the *assertion* that *P* or *Q*? Not, notice, the assertion that *P* or the assertion that *Q* since it is perfectly possible to assert that *P* or *Q* and *not* to assert either that *P* or that *Q*. A plausible line on this, in the case of assertion, is to remember that speech acts paradigmatically involve *speech*; i.e. that asserting that so and so is a performance which characteristically involves the tokening of symbols. And not just *any* old symbols but, specifically, symbols that *mean* that so and so. And the constituency relation for symbols is just the one that the case demands. I.e. while you *can* assert that *P* or *Q* without asserting that *P* or that *Q*, you *cannot* produce a token of the symbol "*P* or *Q*" without producing, inter alia, a token of the symbol *P* and a token of the symbol *Q*.

Similarly, or so one might suppose, in the case of mental states. If believing that *P* or *Q* is being related to a mental symbol then, presumably, it is being related to a *complex* mental symbol; specifically to a complex mental symbol which means that *P* or *Q* and which contains as constituents a mental symbol that means that *P* and a mental symbol that means that *Q*. The postulation of mental symbols thus provides for a straightforward solution to the problem of the constituency structure of mental states, just as it provides for a straightforward solution to the problem of what it is that mental states that have their contents in common have in common.

I don't regard either of the two arguments just rehearsed as decisive (though I do think they're great, whopping straws in the wind.) We'll return to motivating the postulation of mental symbols when we get to the analysis of mental processes. For the moment, however, let's put questions of motivation to one side and concentrate on what a cognitive psychology looks like on the double reduction account; for this is the account which makes the notion of mental representation central, and it is the one that cognitive science has endorsed.

I remarked above the cognitive science, with its emphasis upon the mental representation contruct, is plausibly viewed as continuing a tradition of theorizing about the mind that has its roots in Classical epistemology. In fact, I know of no better way to see how representational theories of the mind work than by considering the relatively simple versions that were developed by these early theorists. One can then understand the current theory as an elaboration of its predecessors, including various wrinkles adopted to avoid difficulties they were beset with. So what I propose to do now is, very briefly and without undue regard for historical accuracy, to have a look at a prototypical early form of mental representation psychology. I will speak of it as "Hume's Theory," though that is perhaps a libel on Hume.

The heart of Hume's psychology is the postulation of a class of mental particulars that he calls "ideas." Ideas are mental symbols *par excellence*. They do for Hume just what mental representations are supposed to do for cognitive science: That is, they provide the source of the intentionality of mental states (in virtue of their semantic properties) and they provide the vehicles of mental processes (in virtue of their causal properties). Let's see how Hume works this.

On Hume's account, the various mental states are identified with correspondingly various ways of entertaining Ideas; so that, for example, to believe that so and so is to entertain the Idea of so and so, and for certain further conditions to be satisfied. It is these "further" conditions that distinguish believing from other intentional states such as, for example, desiring, hoping, intending, doubting . . . etc. The details of Hume's account of these distinctions are notoriously not very convincing and I won't bother to describe them to you. Suffice it, for our purposes, that each intentional state is identified with some or other relation to an Idea that expresses the intentional content of the states. This is, in effect, stage two of the double reduction that starts with linguistic symbols and ends with mental representations. (The first stage is easy for Hume: words—of English, say—get their meanings by becoming Associated with mental states.)

Because intentional states are analyzed as relations to Ideas, and because the intentionality of mental states is exhaustively reducible to the intentionality of Ideas, Hume can, in effect localize the problem of "aboutness" in the following question: what is the property of Ideas in virtue of which they are about things? Hume has an answer to this question: indeed, one of striking simplicity. Ideas are mental *images*; and the intentionality of images is, as it were, inherent; an image, mental or otherwise, is about what it resembles.

It is now widely (and correctly) maintained that this last move won't work. But never mind that for the moment. We can admire the elegance of Hume's theoretical architecture even as we admit that there is something gone badly wrong in the basement. The architectural achievement is this: not only has Hume *localized* the problem of intentionality (by reducing all questions about intentional content to questions about the intentionality of Ideas) but he has provided what one nowadays calls a "naturalistic" theory of intentionality. Intentionality reduces to resemblance, and resemblance (for Hume, anyhow) *is not itself an intentional relation*. Roughly, it's a geometrical relation, and as such it is neither specifically mental nor specifically symbolic. So now one sees another part of the motivation for the double reduction strategy: having reduced linguistic intentionality to the intentionality of mental states, and the intentionality of mental states to the intentionality of mental representations, one then seeks a reduction of the intentionality of mental representations to some complex of *non*intentional relations, *thereby showing that intentionality is not a fundamental feature of the world.*

Thus Hume on intentional states. What about Hume on mental processes? The governing principle here was that mental processes are simply causal interactions among Ideas. In particular, Ideas enter into certain dynamical relations in virtue of being subsumed by the Associative Principle. The history of psychology includes a long (and, on the whole, boring) dispute about how, precisely, the Associative Principle should be formulated. But to all intents and purposes, it amounts to the claim that Ideas that have coocurred in the past will tend to summon one another in the future.

So much, then, for Hume's psychology. It's edifying to contemplate because it shows us very clearly what it's like to have a theory of mind that is grounded in a mental representation construct. But it's also edifying because it shows us very clearly what it's like to have such a theory not work, and we would be ill-advised to ignore the magnitude of the failure. I want to mention three respects in which Hume's vision of cognitive science failed, and then to say something about how the modern variety hopes, in each of these respects, to do better.

First, as I remarked above, Hume's explanation of the intentionality of mental symbols won't do. Hume took it for granted that symbols refer to what they resemble but, as a long line of modern philosophers have pointed out, the resemblance theory of aboutness isn't true. To cite a standard example: King Charles the Xth has his portrait painted; the painting is indifferent and turns out to bear a closer resemblance to Charles' younger brother William that it does to Charles himself; indeed, of William—though not of Charles—the portrait is a spitting image. Still, it's Charles, and not William who is the intentional object of the portrait (i.e. it's Charles that it's a portrait *of*). There is, I think, no doubt of this. Images aren't like Cinderella's slipper; you don't find out whose they are by finding out who they fit best.

Item: if the architecture of Hume's theory if to be preserved, we'll have to have a better account of the intentionality of mental symbols than the resemblance story affords. This is true *even if Hume is right and mental representations are pictures* since, as we've just seen, the resemblance story doesn't work *even* for pictures.

The second major difficulty with Hume's psychology is that associationism doesn't work as a theory of mental process. The usual ground for this claim is just that, in point of fact, the succession of Ideas in thought fails to obey any version of the Associative Principle that psychologists have been able to formulate. But I want to press a rather different point.

When we discussed the role of the intentional states in explanations of behavior, we emphasized the significance of relations among the *contents* of such states for determining their causal interactions; what makes belief/desire explanations work is precisely the content relations among the beliefs and desires. (So "Sutton wanted to knock over the bank because he believed that that was where they kept the money" is OK as a belief/desire theory of Sutton's behavior, but "Sutton wanted to knock over the bank because he believed that two is a prime number" is implausible.) This suggests that the account we want is one which *shows how the causal processes that mental states enter into could respect the contents of those states*. Considered in light of this desideratum, however, association is a terrible basis for a theory of mental processes. This is because associative relations are in an important sense *arbitrary*.

Any mental state can associate with any other, regardless of their contents, so long as the right spatio-temporal propinquities hold among their tokenings. Notice, in particular, that the standard principle of association says nothing about the *contents* of Ideas; it just says that Ideas that co-occurred in the past will tend to associate in the future. Since the content of an Idea does not constrain the associations that it enters into, any Idea can, in principle, have any causal role *regardless of its content*. Not good.

Item: if we're going to preserve the architecture of Hume's psychology, we will need a better notion of mental process than Associationism provides for. In particular, we need some account of how the causal relations among mental states could be determined by corresponding relations among their contents.

Finally, a relatively technical problem: Hume's psychology has trouble with the *productivity* of mental states, and this is not independent of Hume's view of the nature of mental representations.

In saying that the mind is productive, I mean just that the thoughts one actually entertains in the course of a mental life comprise a relatively unsystematic subset drawn from a vastly larger variety of thoughts that one could have entertained had an occasion for them arisen. For example, it has probably never occurred to you before that no grass grows on kangaroos. But once your attention is drawn to the point, it's an idea that you are quite capable of entertaining; one which, in fact, you are probably inclined to endorse.

How *could* mental states be productive in this sense? A natural suggestion is that the productivity of thoughts is like the productivity of natural languages; i.e. that there is an infinity of possible thoughts for much the same reason that there is an infinity of possible sentences. Fine, but how do natural languages manage to be productive? Here the outlines of an answer seem relatively clear. Sentences (and also certain complex words) have internal structure, and the meaning of a sentence is determined, in a regular way, by the internal structure that it has. You can build up new sentences by operations on old ones, and the new sentences will be intelligible because the structure-to-meaning relations that work for the old sentences work for the new ones too. Simple example: a MISSILE is an engine of destruction; an ANTI-MISSILE is an engine of destruction for destroying engines of destruction; an ANTI-ANTI-MISSILE is an engine of destruction for destroying engines of destruction for destroying engines of destruction . . . and so on for as long as the national economy can stand it.

We have, in fact, a rather well worked out theory of how aspects of the meanings of certain relatively complex sentences are determined by their structures and, in particular, by their structural relations to relatively simple sentences that occur as their constituents. This theory goes under the name of logic, and it provides us with striking examples of how it is possible to construct productive representational systems with considerable expressive power form comparatively exiguous primitive bases. The trouble is that, as far as anyone knows, none of the symbol systems that have the combination of productivity and expressive power that mental representations would need to have to be the vehicles of thought are at all like pictures. All the productive symbols systems we know about tend to be organized in very much the ways that sentences are for the simple reason that sentences (unlike pictures) exhibit an orderly relation between the meaning of a whole symbol and the meanings of its parts.

Item: if we are to save Hume's architecture, we will need a better theory of the structure of mental symbols than Hume offers. At a minimum, we will need a theory that is compatible with the productivity of thought.

So much, then, for an early version of mental representation theory. We turn at last to mental representation theory in cognitive science. What I propose to do, in particular, is to suggest what resources cognitive science offers to remedy the three defects we noticed in Hume's account.

Well, to begin with the easiest bit: we had it against Hume's treatment that the system of mental representations needs to be productive in a way that picture-like symbols offer little prospect of being. This problem, however, solves itself since, as I remarked, we have good examples, both from logical formalisms and from natural languages, of symbol systems that *are* productive; and we even have some understanding, in the case of such systems, of the mechanisms whereby their productivity is achieved. These systems do, of course, suffer from being nonimagistic; and that would be a serious defect if

Hume's theory of the intentionality of mental representations (to which their imagistic character was essential) actually worked. But it doesn't, so the assumption that Ideas are like pictures isn't, as things now stand, buying us anything at all. The obvious proposal is to scuttle that assumption and suppose that mental representations are, as the fashionable terminology now has it "discursive"; in effect, sentence-like. So here is the first respect in which cognitive science departs from classical versions of mental representation theory; where they had a mental picture gallery, we have a language of thought.

Actually, and rather gratifyingly, this move to language-like mental representations helps with the second of Hume's problems too. You'll remember that what we wanted, and what Associationism declined to provide, was a theory of mental processes that explains how the causal relations among mental states could somehow connect with their intentional contents. Recent philosophers have, in general, not been very much concerned with mental processes; perhaps this is because philosophical interest in the mind is often epistemologically motivated and knowledge is not a process but a state. Or perhaps it is because so many modern philosophers prove to be behaviorists when closely scrutinized. If you believe that mental states are fictional, you won't want to spend much time on the elucidation of their causal interactions. Anyhow, since mental processes presumably *are* causal sequences of mental states, it is patent that our account of the former must constrain our account of the latter. We're about to see that the need to explain how causal relations among beliefs and desires could reflect content relations among them is quite a lot of what prompts the postulation of mental representations.

The basic pattern of the explanation goes like this. First, let's consider what it is for a mental state to *have* intentional content. This question is, perhaps, the hardest that a theory of mind has to face; but it's plausible that part of the answer lies in the connection between the intentional content of a mental state and its *inferential role.* This idea is most immediately plausible in its application to certain of the logical aspects of intentional content. So, consider what it is for the content of a belief to be conjunctive; eg. for a belief to have the content that *it's raining and the streets are wet.* The hard parts are the "it's raining" and the "the streets are wet" bits; but at least it seems clear enough how to cope with the "and." The *entire* function of the "and" lies in bestowing a certain inferential role on the belief as a whole; viz. in determining that the belief is inferable from the belief that it's raining and the belief that the streets are wet; and that, conversely, each of these latter beliefs are inferable from it. If this doesn't strike you as self evident, compare the belief that it's raining *and* the streets are wet with the belief that it's raining *or* the streets are wet. The difference between the content of these beliefs seems to be precisely captured by a certain difference in their inferential roles; viz. that the conjunctive belief it inferable from both of its conjuncts while the disjunctive belief is inferable from either of its disjuncts.

Intentional content and inferential role seem, in short, to be very intimately related. Some philosophers think that they're identical, but I don't; and we needn't assume anything that strong for the purposes at hand. We needn't even assume that inferential role is an *ingredient* of content, so long as we're prepared to admit that similarity of content goes along with similarity of inferential role in an important number of cases. This last seems to be patently true, and it is enough to assure us that, in this important number of cases, you could get the causal role of a belief to reflect its *content* if only you could get the causal role of a belief to reflect its *inferential* role.

But how on earth are we going to do *that*? Inferential role is, after all, itself a semantic notion rather than a causal one; the fact that you can infer (demonstratively or inductively) from one belief to another doesn't, in and of itself, have any *causal* implications at all. On the contrary, to say that you can infer from P to Q is just to say that if P is true then Q is true.

So there are three balls in the air: intentional content, inferential role, and causal role. What we want is a theory of mental processes which provides a mechanism for the parallelism between causal role and intentional content. We could get such a mechanism if we could see how the inferential role of a mental state could determine its causal role since—on some views—inferential role is what the notion of intentional content unpacks into, and—on any view-inferential role and content run together in very many interesting cases. But what kind of mechanism could serve to establish the required parallelism between the causal role of a mental state and its inferential role?

This is a kind of problem that has a solution for *symbols;* specifically, for objects that have internal syntactic structure (of the sort, by the way, that we relied on for the solution of the productivity problem). The idea is this: the *inferential* relations among symbols are inherently semantic and have nothing directly to do with their causal roles. But the *syntax* of a symbol *is* very closely related to certain of the physical properties which determine its causal role, such as its geometrical shape. (When I say, for example, that symbols have internal syntactic structure, I mean that some symbols literally contain others as *physical parts;* in the way that the symbol "it's raining and the streets are wet" contains the symbol "the streets are wet" as a part. Since it is such physical properties of symbols as their shape that determines their syntax, it's fairly easy to see how there could be environments in which *the causal role* of a symbol correlates with is syntax. It's easy, that is to say, to imagine symbols interacting causally in virtue of their syntactic structures; the syntax of a symbol might determine its causal role in much the way that the geometry of a key determines which locks it will open.

And now, finally, we come to the point: it's a major insight of modern logic that certain of the inferential relations among symbols can be, as it were, mimicked by their syntactic relations. For example, within certain limits, the *semantic* relation that holds between two symbols when one is *deducible* from the other can be mimicked by *syntactic* relations in virtue of which one of the symbols is *derivable* from the other. We can even build a machine which has, within certain limits, the following property: the operations of the machine consist entirely of transformations of symbols; in the course of performing these operations, the machine is sensitive *solely* to syntactic properties of the symbols; and the operations it performs on the symbols are entirely confined to alterations of their shapes. Yet the machine is so devised that it will transform one symbol into another if and only if the symbols so transformed stand in a certain *semantic* relation; viz. if and only if the one symbol entails the other. Such machines are called computers.

So now a number of points emerge. In the first place, we see the point of the cognitive scientist's frequent insistence on the analogy between minds and computers. Computers are a solution to the problem of mediating between the casual properties of symbols and their semantic properties, so that *if* the mind is a sort of computer, we begin to see how you can have a theory of mental processes which succeeds where the associative theory failed; viz. a theory which explains how there can be nonarbitrary content relations among causally related beliefs and desires.

But, second, we see why there have to be mental representations if the cognitive science proposal is going to work. In computer design, causal role is brought into phase with content by exploiting parallelisms between the syntactic properties of symbols and their semantic properties. But, of course, that idea won't do the theory of *mind* any good unless there are *mental* symbols; i.e. mental particulars simultaneously possessed of semantic *and syntactic* properties. Treating mental states as relations to symbols—hence as relations to syntactically structured objects—is the only way anyone has yet thought of to explain how the causal properties of beliefs and desires could connect with their contents. There must be mental symbols because, in a nutshell, only symbols have syntax and our best available theory of mental processes needs the picture of the mind as a syntax-driven machine.

What about Hume's third problem—viz. the failure of the resemblance theory to account for relations between Ideas and what they are ideas *of*? What's wanted is a theory of the relation between symbols (specifically, mental ones) and the nonliguistic objects which they stand for. And, as you'll remember, there is a certain condition on that theory; viz. that it be "naturalistic" in the sense that it reconstruct the semantic properties of mental symbols without appeal either to *other* semantic properties or to mental ones.

Roughly speaking, there have only ever been two proposals for a naturalistic theory of the relation between symbols and the things they symbolize. One is that the relation is somehow based on resemblance (mental symbols are mental pictures and they stand for what they look like) and the other is that it's based on causation (mental symbols are discursive, and they stand for what causes you to entertain them). Since we know that the picture story is no good, it follows that some version of the causal story must be true. There are, however, certain difficulties with the causal story; in particular, that nobody knows exactly how to tell it. We are working on it at this very moment.

That ends our survey. What we've got is, clearly, work in progress. You can view it as an attempt at urban redevelopment. We want to save a theoretical architecture we inherited from Hume and his friends; to shore the ancient building up with some fancy modern engineering. I think the resulting structure is pretty persuasive; the basic fact about mental states—their peculiar admixture of intentionality and causality—promises to fall out nicely from an account of the mind as essentially a symbol processing mechanism. And, though I haven't time to set out the details, it's worth emphasizing that there are some; there exist fairly well developed theories of the mental processes that underlie certain of our cognitive capacities (particularly in the areas of language and visual perception) and these theories have met with an appreciable degree of empirical success. There are psychological facts that we can explain now that we couldn't explain before; and the explanations of these facts turn on detailed hypotheses about how mental representations are structured and how their structure determines the course of mental processes.

Coda: Deliniating "the place of values in a world of facts" is the characteristic philosophical worry of our time. Admittedly, when people have this worry it isn't usually *truth* values that they have in mind. But it ought to be. Semantical and intentional properties are at the heart of the propositional attitudes, and only a psychology that takes propositional attitudes seriously gives the slightest promise of providing theories of behavior adequate to the complexity of human action. And problems about human action *are* what people who worry about values worry about.

To take the attitudes seriously is to view them as *causes* of behavior. But the problem is that it's hard to see how semantical and intentional properties *could* belong to the causal order. At a minimum, if we are to save our basic intuitions about human action, there had better be something like intentional causation. However, if we're also to save our scientific metaphysics, it had better turn out that the mechanisms of intentional causation are not *sui generis* (which is just to say that, at a minimum, they had better turn out to be *mechanisms*). It is, however, no use insisting that "at a minimum they'd better be" if, in fact, you have no idea of how they could be. Until quite recently, we have had no idea of how they could be.

This is a real metaphysical tension; one that comes within hailing distance of the traditional fact/value antinomies. So what's most interesting about the state of play in cognitive science, viewed in this light, is the possibility that one of these fact/value problems may actually be in the process of getting solved. It is, no doubt, a catch as catch can sort of a solution; intentional causation is at best only approximated, and what it's approximated by is syntactical rather than semantic. But perhaps that's not so bad; intentional causation *had* to turn out to be really something else. It was never in the cards that intentionality would prove to be a *fundamental* feature of the world (in the way that spin, or charge, or charm may turn out to be fundamental features of the world.) The real worry was that maybe the behaviorists were right and it would turn out to be *nothing at all.* That worry, however, seems increasingly to have been unfounded.

There is, in short, a distinct prospect that we'll come out of all this with a fair number of our intuitions intact; with more of our "freedom and dignity," even, than had quite recently seemed likely. If that isn't all that Humanism demands, it is at least more than Humanism has been accustomed to getting lately. Proving, I suppose, that e.e. cummings was right after all: "Nobody loses all of the time." Not even us.

NOTES

1. That is, psychological explanations in which the behavior of an organism is accounted for by reference to its beliefs and desires. See below.

2. Some people (I among them) don't like *inscriptions* to be truth-valuable. But the present point—that inscriptions have intentional/semantic properties as well as physical/causal ones—is still intact. At a minimum, inscriptions express propositions, and that's a semantic property within the meaning of the act.

TWO CONTROVERSIES IN EVOLUTIONARY BIOLOGY: SALTATIONISM AND CLADISM

MARIO BUNGE

I. Introduction

According to Karl Popper and his followers, criticism is the mark of rationality and, in particular, of science. On the other hand, according to Thomas Kuhn and his followers, controversy is the mark of ideology. There is a grain of truth in each of these opinions. Indeed, although controversy is an important component of scientific research, before an idea or a procedure can be subjected to critical examination and debate it must have been created. The engine of science is creation: discovery or invention. Criticism is the regular-tor of the engine.

Let us have a look at the nature and role of scientific controversy. Evolutionary biology has been controversial from birth. It has been attacked from the outside—in particular by religionists—and it has not ceased to evolve as a result of research. This evolution has not always been smooth. Thus, the synthetic theory of evolution was revolutionary, and some of the current debates within the field are vigorous and at times acrimonious. Just as with ideological matters, most people tend to rally around one or the other position, without giving much thought to the methodological and philosophical problems involved, while the opponents of evolution watch these infightings with glee.

This paper will examine two current controversies in evolutionary biology, in an attempt to identify and clarify some of the philosophical and methodological issues involved in them. Our main conclusions will be that (*a*) the debates are likely to remain inconclusive as long as the philosophical and methodological problems are not settled, and (*b*) the creationists can derive no comfort from such infightings.

II. Saltationism vs. gradualism

Throughout the history of ideas there has been a tension between the partisans of continuity and those of discontinuity. Aristotle and Leibniz championed the former, whereas the atomists and the theologians—strange bedfellows though they be—stressed discontinuity. The controversy has not been confined to metaphysics: it has involved conflicting world views, and even religion and politics. Naturalistic thinkers, as well as political liberals and reformists, usually espouse Leibniz's "law of continuity," according to which all changes proceed gradually. On the other hand believers in miracles and partisans of political revolutions usually stress discontinuity. Lyell in geology, and Darwin in biology, espoused gradualism in their naturalistic accounts of the history of our planet and of life on it. Now that evolutionary biology is firmly established—in science if not in pop culture—it is possible to take sides in the controversy without being accused of doing so with theological or political motivations. But it also becomes possible to advocate a synthesis of the two views in certain cases.

Gradualism suffered several setbacks, not only in atomic physics but also in biology. A first setback was when genetics disproved the blending or soft theory of heredity and replaced it with the particulate or hard theory of genic change. A second setback occurred only two decades ago, when plate tectonic theory showed that continents could break up or join together, thus suddenly opening up or closing down thousands of ecological niches, which could explain massive species extinctions or speciations.

The latest attack on gradualism is the theory of punctuated equilibria (Eldredge and Gould 1972, Gould and Eldredge 1977, Gould 1980). According to this theory every evolutionary line is most of the time uneventful (stasis), but occasionally there are very quick jumps ("punctuations"). Moreover, such jumps would not be the result of ecological changes, but of developmental (in particular embryological) constraints or accidents. We shall argue that neither of these challenges poses a serious threat to the neo-Darwinian theory, although all of them can enrich it. Let us examine them.

Strict genetic gradualism is false because differences between individuals and their offspring, however small, are not arbitrarily small; and, if accumulated over a large number of generations, they are detectable. The same holds, a fortiori, for differences between species. Thus if one of the morphological differences between two annual plant species is 1 cm in the length of some trait, and this difference took 1 million years to evolve, then each successive generation contributed on the average one-hundredth of a micron. So, it is not true that *natura non facit saltus*. However, this does not prove that a new species can emerge in one generation, except perhaps by hybridization.

As for plate tectonic theory and its companion, continental drift, they explain certain sudden changes, some of which have been massive. For example, when Panama joined with South America only a few million years ago, many of the North American carnivores migrated southward and eliminated some of the huge South American herbivores. An even more dramatic mass extinction occurred about 225 million years ago at the end of the Permian period, when about half of the animal species became extinct. This well known event is often explained as a result of the coalescing of all the continents (i.e. the formation of Pangaea), as a consequence of which the total area of shallow water, where most of the animals lived at the time, was roughly halved. Another massive extinction occurred only 65 million years ago at the end of the Cretaceous. (A current popular explanation is again a mechanismic one: meteoritic impact.) However, the catastrophe was not instantaneous. In particular, its most famous episode, that of dinosaur extinction, was a gradual process lasting at least 7 million years and likely to have been caused at least partly by competition with immigrating ungulates (Sloan *et al.* 1986).

We have mentioned only externalist, and moreover catastrophic, explanations of mass extinctions, because we are discussing

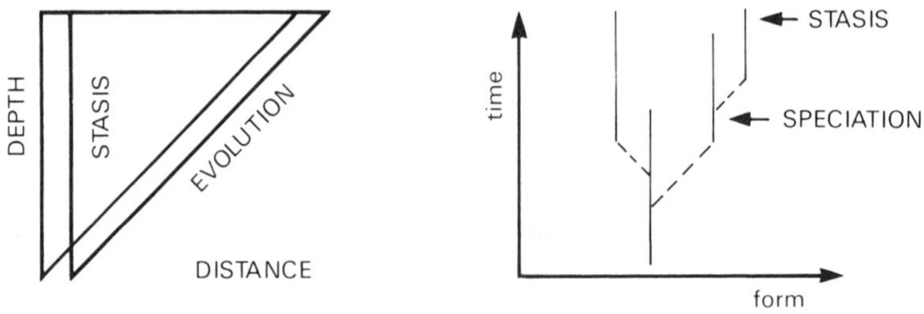

Figure 1. Punctuated equilibrium hypothesis. (*a*) Fossil quarries. (*b*) Statis-speciation sequences.

discontinuity. However, there are also internalist explanations , such as the propagation of nonadaptive genes, which can be rather swift in small populations. However, the most likely, or at least the most common, extinction mechanism may have been a combination of internal and external changes. One such hypothesis is that monomorphism—very small genic variability within a given species, as in the case of the Cheetah—combined with niche perturbation, is likely to decimate populations to extinction point (O'Brien *et al.* 1983).

As for the punctuated equilibria hypothesis, it is a gallant effort to abstain from using the incompleteness of the fossil record as an excuse or a way to protect hypotheses from refutation. Moreover, it is an attempt to turn lack of positive evidence for gradualism into positive evidence for saltationism. The proponents of the hypothesis read the fossil record in a literal manner, and assume that the history of a species (actually a population) in a given locality is roughly one of stasis. When speciation does occur it takes place in a different area and it consists in the splitting of lineages as a consequence of adaptation to a new environment (i.e. allopatric speciation). Hence discontinuities are to be sought, not in a given vertical rock column, but along some incline: see Fig. 1. Such punctuations would be abrupt and could result from macromutations producing deviant individuals preadapted to their new environment. This latter conten tion has been softened in the latest version. However, the main claim, that there is macroevolution rather than gradual evolution, has been kept.

Like any other scientific hypothesis, the one under consideration should be evaluated in the light of both empirical evidence and theory. Now, the former, which involves mainly vertical stratigraphic sequences, is ambiguous: while some data (e.g. Williamson 1981 and Iltis 1983), favor it, other (e.g. Cronin 1981) undermine it. Moreover, the Cretaceous Bivalvia, perhaps the best studied group, and one that includes over 100 lingeages of a large morphological and ecological variety,

> reveal a vast array of evolutionary patterns, representing broad variations in evolutionary data [. . .] which may range from gradualistic [. . .] through mosaic or reticular [. . .] to punctuated or stepwise in nature [. . .] Evolutionary rates range from extremely rapid [. . .] to extremely slow [. . .] Further, evolutionary rates may change dramatically *within* single lineages subjected to varying stress levels through time, and vary widely *between* co-occurring lineages subjected to similar sets of changing environmental parameters (Kauffman 1978 p. 277).

Finally, an analysis of a 10-million-year lineage of planktonic foraminifera suggests that evolutionary change is sometimes very rapid, though never abrupt: this hypothesis has been dubbed *punctuated gradualism* (Malmgren, Berggren and Lohrmann 1984). In short, only *some* phylogenies are of the stasis-explosion type. Gould (1983) himself has admitted this.

As for theory, the situation seems to be this. The standard theory of evolution does *not* postulate that all evolution is gradual, let alone that all evolutionary processes proceed at a constant rate (Maynard Smith 1982). In principle both moderate speeds (on the order of one new species per million years) and very large ones (on the order of one per 50,000 years) can and do occur, depending on the intrinsic rate of mutation, environmental changes, and other factors. So, the hypothesis that there are quick evolutionary changes is compatible with standard theory, which, being extremely general, abstains from assuming any definite rates. Moreover, the hypothesis had been put forth by one of the builders of the synthetic theory, namely Mayr (1970, 1982b).

Discontinuities, or rather extremely rapid changes, are not only compatible with the standard theory but they can be explained, though perhaps not predicted, by it. In fact, a number of mechanisms have been proposed for "discontinuities," such as colonization of a new habitat, change in function of an organ, selective mating, and (in small populations) random genetic drift. Either of these mechanisms, acting over thousands of years, can produce the *appearance* of a sudden change or macromutation. It is all a matter of *tempo*, and the latter is determined by the underlying mechanism. As Sewall Wright (1982), the doyen of American evolutionists, put it, there is microevolution and macroevolution: the former consists in improved adaptation to a particular niche, whereas the occupation of a vacant niche favors macroevolution. In both cases there are of course genic changes, but there is no such thing as macromutation. The rapidity of macroevolution results from new opportunities and the near absence of competition in the new niche. In other words, whereas microevolution proceeds mainly by descent with modification and selection, macroevolution is made possible by a drastically diminished selection pressure combined with random variations (Newman *et al.* 1985).

In sum, the standard theory of evolution is consistent with macroevolution provided the latter be interpreted as tachyevolution or very fast evolution—which of course is not the way the punctuationalists see it. Consequently the punctuated equilibrium hypothesis is far from being revolutionary (Lande 1980, Stebbins and Ayala 1981, Jones 1981, Mayr 1982b). What *is* novel is the hypothesis, originally proposed by Goldschmidt in 1940, and intially favored by the champions of punctuated equilibrium, that tachyevolution is "decoupled from microevolution" for being the product of macromutations resulting in "hopeful monsters." But genetics holds out no hope for such monsters. In fact it is well known that any morphological changes, when viable at all, are comparatively slow and polygenic (i.e. involving many genes). On the other hand mutations with large effects and involving only a few genes, though possible, are improbable and usually deleterious. So, the strong version of the punctuated equilbrium hypothesis seems to be incompatible with genetics. But this point is moot and philosophically uninteresting.

On the other hand the following points are of great methodological importance. One is that the absence of evidence for a hypothesis (e.g. gradualism) does not constitute positive evidence for its negation (e.g. saltationism). Thus the gaps in the fossil record do not confirm saltationism, the more so since they can often be explained by taphonomy (the science of the evolution of fossil traces) together with the haphazard nature of paleontological discovery. (The known sites of fossil traces are not statistically random samples of the total fossil population.) The second point is that disconfirming empirical evidence weighs at least as much as favorable evidence. And, as noted a while ago, while *some* evolutionary lines confirm the alternation of quick and slow (or nil) evolution, other disconfirm it. In other words, the hypothesis is at best existential—i.e. it starts with the term "some"—not universal. Hence it can hardly be incorporated into the *general* theory of evolution. (In other words, data about evolutionary rates are just that: data, not postulates.)

Although some creationists have presented the gradualism-saltationism controversy as evidence for the bankruptcy of evo lutionary biology, it is clearly a family quarrel within the latter. Its main virtue is that it has greatly enhanced interest in the crucial problem of the "mode and tempo" of evolution, and has reminded us that the problems of the precise mechanisms of speciation and extinction is the most important single open problem in evolutionary biology.

Progress in grappling with this problem has been somewhat slow for a number of reasons. Firstly, because it has attracted far too few workers—possibly because it is often supposed to have been solved long ago. Secondly, because research on it has been hampered by fuzzy (in particular holistic, as opposed to systemic) thinking (Futuyma 1983). Thirdly, because, although several speciation mechanisms are likely to operate jointly, every biologist tends to focus his attention on his own favorite. (In particular, the importance of mutation and reproductive isolation may have been exaggerated, whereas ecological factors and sexual competition seem to have been underrated.) Fourthly, because the estimates of rates of morphological changes are still quite unreliable. (Typically, laboratory selection experiments yield rates many times greater than those suggested by the fossil record: Gingerich 1983.)

To conclude this section: some evolutionary processes are so quick that, on a geological time scale, they look like big leaps. But such jumps are not more frequent than slow evolutionary processes, and in any event they do not demand a radical revision of the standard theory of evolution. If anything, they demand that a closer attention be paid to taphonomic processes, i.e. to the evolution of fossils. We need a more sophisticated and detailed theory of evolution for other reasons, e.g. to explain speciation and extinction, and to achieve a much more intimate union of population genetics and evolutionary ecology. (See e.g. Roughgarden 1979.)

III. Cladistics vs. standard systematics

The theory of evolution had, predictably, a profound influence on biological systematics. The latter began to be seen as the investigation of kinship relations and as culminating in the drawing of genealogical (or phylogenetic) trees. Moreover, it was hoped that life on Earth originated only once, so that all terrestrial organisms descend ultimately from the original cells. Thus a single genealogical tree with a few million branches—most of them now dry—would encompass in one stroke the long and involved history of life on our planet and the corresponding marvelous variety of life forms and styles.

However, although nearly every systematist nowadays admits that only evolutionary (or phylogenetic) classifications are adequate, and that every classification must match a genealogical tree, there is still considerable controversy over the best way of establishing such classifications—so much so that some cynics hold that there is no such best way, whence systematics is bound to remain largely subjective and closer to art than to science.

I submit that this unsatisfactory state of affairs is due not only to the complexity of the subject but also to certain philosophical confusions. Of these, the main trouble-maker is the holistic-nominalistic view that species are real wholes (concrete individuals), so that being a member of a species is similar to being a part of an organism. This thesis involves a confusion between the set theoretic relation of membership in a set with the ontological relation of part to whole. In turn, this confusion has induced the confusion of species (a collection) with a biopopulation (a concrete system). Unless one forms classes at some point, i.e. unless a low order taxonomic category (taxon) is construed as a class, no classification is possible. If species are construed as individuals, then the first taxon is of the genus level. But if this is done then individuals become parts of species, and the latter members of genera, with the result that individuals belong to no genera. (Bunge 1981, 1985.) The correct logical relation between taxa, such as species and genera, is that of inclusion ($\subseteq$), not membership ($\in$). For example, *Fido* $\in \subset$ *Canis familiaris* $\subset$ *Canidae* $\subset \ldots \subset$ *Chordata*. (See Bunge 1983, 1985.)

Unfortunately some of the very founders of the synthetic theory of evolution, notably Dobzhansky (1970) and Mayr (1982a), fell into the holistic-nominalistic trap as early as in the 1940's. And, being diffident of logic, they paid no attention to the devastating criticisms of Gregg (1950). Moreover, some science writers and some philosophers (mainly Hull 1965, 1974) lent their support to the muddle instead of dissolving it with the help of elementary logic. Thus authority consecrated an elementary logical blunder that has prevented evolutionary taxonomy (though not evolutionary systematics) from advancing.

A first result of confusing biospecies with biopopulations is to mistake *family trees* for *histories in space-time*. The history of any material thing can be diagrammed as a spatio-temporal graph. This applies, in particular, to a biopopulation that splits into two geographically isolated populations: see Fig. 2. In addition to such a space-time diagram, a biopopulation has a genealogy, or phylogram, that can be diagrammed on the diversity-time plane. If the population splits into two different parts, the members of the new populations may continue to belong to the same species, at least for some time: this case of stasis is represented by the straight line in Fig. 2a. But geographical splitting may be accompanied by speciation, i.e. a branching in the diversity-time plane: see the fork in Fig. 2b. This is the case of allopatric speciation. On the other hand, in sympatric speciation there is no spatial divergence between the populations; instead, there is genetic or morphological (inclusive "or") between their members. In sum, genealogical trees must not be mistaken for space-time histories, if only because the former refer to species, which are concepts, whereas histories refer to populations, which are material things.

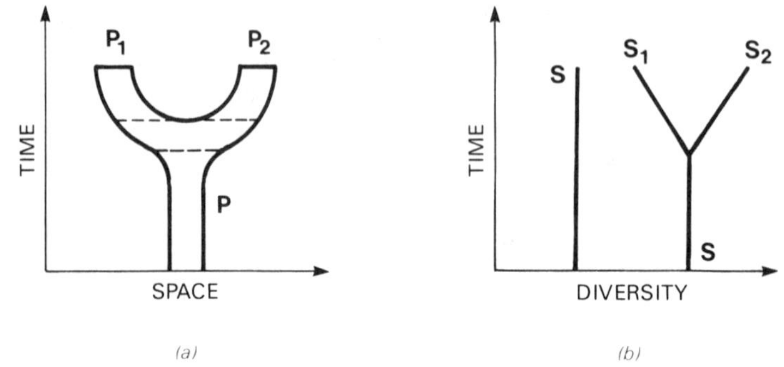

Figure 2. (*a*) A Population P splits into two populations P_1 and P_2 occupying different territories. (*b*) The splitting of the population is accompanied by stasis (single species S) in one case and by speciation (branching out of S into S_1 and S_2) in the other.

The confusion we have just noted is associated with another, namely the view that defining a species by the characters shared by its members results in forming "static" species concepts akin to Platonic or Aristotelian "essentialistic" definitions (Hull 1965). This view is mistaken because it is impossible to trace the history of a thing X unless one has formed *some* idea of X—which prior idea will probably be enriched and refined as the investigation of the history of X proceeds. Still, the holist-nominalist is likely to challenge the very applicability of the notion of a kind on the ground that evolution is gradual. Thus he may point to the region between dotted points in Fig. 2a holding that, as long as it is in that slice of spacetime, the branching population is in a transition state, so that its members cannot be said to belong to either the original species S or its daughter species S_1 and S_2. In short, the very concept of a species would founder during the most interesting episodes of evolution, namely those of speciation. Clearly, the concept of evolution, which is essentially dynamic, would continue to be useful where that of species does not work.

Again, this holistic thesis derives from confusing things with collections and, therefore, population diagrams with genealogical trees. Even assuming that evolution is gradual, though admittedly quick at times (Sect. III), it is not continuous like motion, because

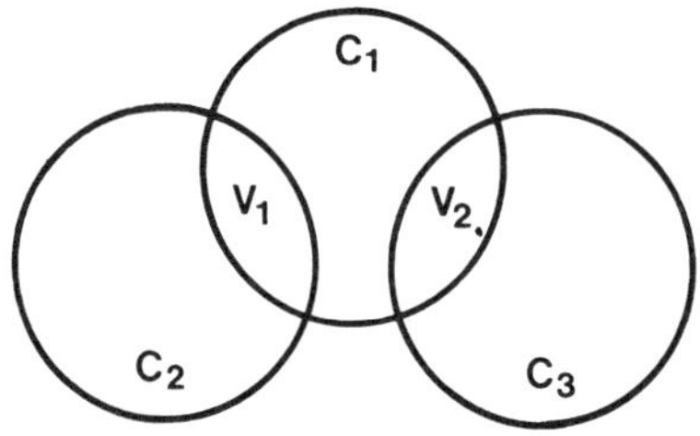

Figure 3. Partially overlapping sets of properties are consistent with disjoint collections of individual organisms possessing those properties. Example: variety V_1 is constituted by organisms with properties P_1 (determining class C_1) and P_2 (corresponding to class C_2); variety V_2 is composed by individuals with properties P_1 and P_3. The collections V_1 and V_2 are mutually disjoint.

genic changes are abrupt. Indeed such changes result in qualitatively different DNA molecules and consequently in a difference in the protein composition of the organism. Therefore it is possible to spot the founders of a new variety or race: breeders do it all the time. However, such a branching into two or more varieties may not show up in the population diagram if the population continues to occupy the same territory.

Whether the new varieties are sufficiently different from one another and from their immediate ancestors to deserve being called *new species*, is beside the point. The point is that gradual evolution is compatible with the grouping of organisms into different and moreover mutually disjoint classes. This applies, in particular, to transition groups such as lungfishes and cynodonts. To claim that the very concept of a species fails us for such transition kinds, is to mistake sets of properties for collections of organisms. The former may overlap partially whereas the latter need not: see Fig. 3.

The attacks on the "static" concept of species, and the attempt to replace it with the evolutionary concept of descent, can only muddle up systematics for, as noted above, before attempting to trace the history of X we must have formed some idea of what X looks like. Once the history of X has been conjectured we may have to revise the original characterization of X—which in turn may force us to alter the phylogenetic hypothesis (Mayr 1981). This feedback process of successive approximations it not a mark of confusion but of science.

Many Darwinian species coincide with the corresponding Linnean ones. The main novelty lies in the relation of descent involved in phylogenetic trees—which trees were of course unknown to the pre-Darwinian biologists. Hence the Darwinian concept D of biospecies boils down to the conjunction of the classical concept C, defined in terms of common characters, and the new condition A of common ancestry, i.e. $D = C \& A$. (Hence the intension or content of D includes that of C, i.e. $\mathcal{I}(D) \equiv \mathcal{I}(C\&A) = \mathcal{I}(C) \cup \mathcal{I}(A) \supset \mathcal{I}(C)$. See Bunge 1974, Ch. 4, on a calculus of intension.) Therefore the claim, sometimes heard, that the two concepts, C and D, are "incommensurable," is false. It is not that evolutionary biology rejects the grouping of organisms by relations of similarity, in particular homology, but that it *supplements* and *explains* such natural groupings in terms of common ancestry.

Exaggerating the inadequacy of any preliminary "static" concept of biospecies, and holding that only the relation of descent should define it, leads to *cladism* (Henning 1966). In fact the main thesis of cladistics is that two groups of organisms are closer, the more recently they have diverged from their common ancestors. Therefore cladists regard the nodes or branching points in a genealogical tree as more important than the lines joining them. In other words, they recommend that the study of descent precede that of diversity. For this reason they tend to attribute themselves the monopoly of *phylogenetic* systematics, calling "traditional" the biologists who employ systematically the synthetic theory without using the particular postulates of cladism (Lipscomb 1984).

Cladistics includes also three mutually independent hypotheses: that (*a*) the simpler of two hypothetical genealogical trees (cladograms) is always to be preferred; (*b*) all branchings (specifications) are binary, and (*c*) the rate of divergence between any two species is the same (i.e. time is the only independent variable in evolution). Cladistics has been the subject of a spirited controversy over the past few years, and its partisans are enthusiastic, vocal, and well organized, having banded together in the Willi Hennig Society. We proceed to show what is wrong with it.

First of all we must raise a *logical* objection to the thesis that diagramming must precede classing: one must form a family of classes before he can proceed to ordering them. Try to write, say, "$S_1 < S_2 < S_3$," where '<' stands for the relation of descent, without having the least idea as to what the S's stand for. Of course once a serial order has been conjectured on the basis of similarities, primitive (or derived) traits, vestigial organs, or what have you, it becomes possible to take a second look at the preliminary family of classes—e.g. in order to discover homologies or evolutionary parallels. As noted before, the classing-diagramming process is one of negative feedback and therefore of successive approximations, hence an open-ended one. But the point is that one cannot *start* by investigating relations of descent. Descent is not *the* defining characteristic of a group of organisms but the clue to the relations between species and the higher order taxa. We shall return to this point in a moment.

Having disposed of the cladistic thesis that the reconstruction of phylogenies (the conjecturing of cladograms) should precede everything else in systematics, let us take a quick look at the three subsidiary theses of the cladistic strategy mentioned at the beginning. The parsimony thesis is both ontologically and epistemologically indefensible. The first because it is simply not true that nature always chooses the simplest paths—as it ought to if it proceeded teleologically. And epistemological parsimony is suspect because, among other reasons, the history of science suggests that we advance knowledge more often by complicating than by simplifying. (For more on the myth of simplicity see Bunge 1963.)

As for the thesis that all branchings (speciations) are binary, it is a nice simplifying assumption that may hold in many, perhaps most, cases. However, there are counterexamples to it. Nor is the hypothesis of the constant rate of divergence true (as we saw tacitly in Sect. III). Given the differences in genic diversity and in the environment, it would be miraculous if such regularity held. In any case, the cladograms constructed on the basis of the main hypotheses of cladistics are sometimes inadequate. For example, cladists place crocodiles closer to birds than to lizards, which is quaint because crocodiles and lizards belong to the class of reptiles.

Descent, then, is necessary but insufficient to reconstruct genealogies. Two phyletic lines may have diverged as recently as 50,000 years ago, and yet they may have become quite different for having adapted to very different environments or for some other cause. That is, though genetically very close, two species many be morphologically quite distant—or conversely. Chimp and man are a case in point: though genetically very close, they are enormously different at the organ level (particularly in their brains) as well as ecologically.

The taxonomic method that employs *both* descent and morphological (and genetic) diversity is, of course, standard ("classical") evolutionary taxonomy. Moreover, in this taxonomy—unlike in phenetics—the sheer number of similarities is not as important as the similarities that have an *evolutionary raison d'être*, i.e. the homologies. (Two organs or "structures" are said to be *homologous* if and only if, in addition to discharging similar functions, e.g. swimming, they have a common ancestry. Thus the flippers of the seal and the manatee are homologous: both are adaptations of legs to marine life. On the other hand those flippers are only analogous to the fins of the fish.)

In sum, standard taxonomy heeds Julian Huxley's advice: Characterize each group of organisms as a unit constituted by entities with a number of characters in common (a *grade*), *and* as a class of relatives (a *clade*). In this way the logical relation of inclusion between taxa of different levels or oders is explained in terms of "propinquity of descent," to use Darwin's phrase. See Fig. 4.

Normally descent is not observable, so it must be conjectured on the basis of morphologically (and if possible also genetic) closeness. For example, we hypothesize that humans, dogs, bats and seals have common ancestors, because they are all mammals and they all share, among other features, the same gross arrangement (anatomical "plan") of bones in the forelimbs. But their being mammals is not enough to make them close relatives, for it is possible that mammals evolved more than once. Nor does a common anatomical "plan" of some "structure" suffice, for we must always reckon with the possibility of convergent evolution. (The strong point of pheneticists is to remind us of such precautions.)

The basic inference pattern in guessing common descent is this:

Common ancestry implies morphological closeness.
Now, species S_1 and S_2 are morphologically close.
Ergo, maybe S_1 and S_2 have a common ancestry.

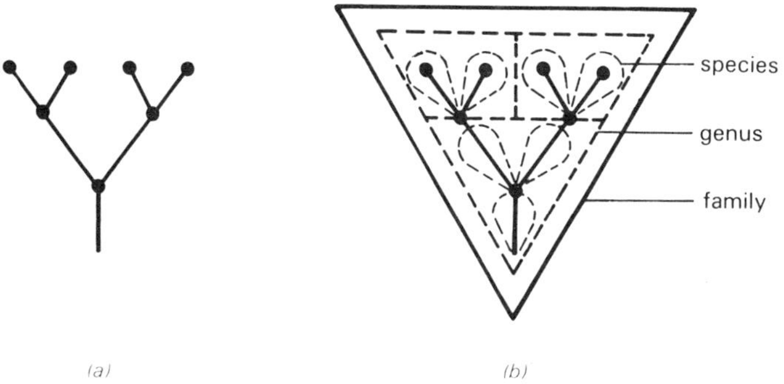

Figure 4. Relation between the logical relation of inclusion of groups of organisms into wider groups, and the biological relation of descent. In (*a*) a phylogenetic tree is hypothesized; in (*b*) an evolutionary classification is proposed on the basis of both morphological similarity and common descent (i.e. genetic kinship). After Simpson 1961.

This is not a valid (deductive) inference but a plausible (seductive) one. Morphological closeness is used in it as an indicator of kinship, not as a foolproof criterion of it. (For an analysis of the notion of an indicator, see Bunge 1983.)

To conclude, all contemporary biologists are agreed that biological evolution has occurred and that it can be summarized as a huge genealogical tree rooted to some primitive cells. What is in question is not the real tree but the best manner of drawing it: not the process but our account of it. This is no easy matter because it involves going beyond the empirical evidence: in fact all phylograms are hypothetical and therefore tentative. For this reason reconstructing family trees is something of an art, just like reconstructing segments of human history, or building scientific theories. But, being a scientific endeavor as well as an art, the diagramming of genealogies is corrigible in the light of new evidence or new biological principles. This process, though convergent, is open-ended: the task of the systematist is never finished. Creationists are likely to take this for a sign of weakness of evolutionary biology. In fact it is a mark of its scientific character: only ideologies claim to be perennial.

IV. Conclusions

In all fields of research and rational action criticism is an indispensable control—though not more important than observation, measurement, or experiment. For this reason criticism, and the controversies it generates from time to time, are necessary in all fields of research and rational action to weed out error. Therefore the occurrence of controversy in science and technology are signs of growth not of decay—provided that not everyone becomes a full time controversialist. The furnace needs a thermostat but this control, though necessary, is not sufficient to keep the house warm. Efficiency calls for a *via media* between the Kuhnians—furnace gone wild without a thermostat—and the Popperians—thermostat gone idle without a furnace to control.

Saltationism and cladism have given rise to controversies in evolutionary biology. Seen in perspective they have not been scientific revolutions but minor heresies. The controversies about them are still going on and they have been fruitful. Unfortunately the controversialists have made no use of elementary methodological tools that would have clarified the issues and spared some mistakes. Therefore they have no legitimate cause for complaint if the philosopher, uninvited and vastly ignorant of substantive details, enters the fray. The only way to keep the meddlesome philosopher away is to eat a healthy philosophical apple a day.[1]

NOTE

1. I take pleasure in acknowledging the useful, though not always heeded, criticisms and comments made by John Maynard Smith (University of Sussex) and Osvaldo A. Reig (Universidad de Buenos Aires).

McGill University

BIBLIOGRAPHY

Bunge, M. (1963) *The Myth of Simplicity*. Englewood Cliffs, N.J. :Prentice-Hall.

Bunge, M. (1974) *Treatise on Basic Philosophy*, Vol. 1: *Sense and Reference*. Dordrecht: Reidel.

Bunge, M. (1981) Biopopulations, not biospecies, are individuals and evolve. *Behavioral and Brain Sciences*, 4:284-285.

Bunge, M. (1983) *Treatise on Basic Philosophy*, Vol. 5: *Exploring the World*. Dordrecht and Boston: Reidel.

Bunge, M. (1985) *Treatise on Basic Philosophy, Vol,. 7, Philosophy of Science & Technology*, Part II: *Life and Social Sciences, and Technology*. Dordrecht and Boston: Reidel.

Cronin, J. E., N. T. Boaz, C. B. Stringer and Y. Rak (1981) "Tempo and mode in hominid evolution." *Nature*, 292: 113-122.

Dobzhansky, Th. (1970) *Genetics of the Evolutionary Process*. New York: Columbia University Press.

Eldredge, N. and S. J. Gould (1972) "Punctuated equilibria: an alternative to phyletic gradualism." In T. J. M. Schopf (ed.), *Models in Paleobiology*, pp. 82-115. San Francisco: Freeman, Cooper & Co.

Futuyma, D. J. (1983) Review of C. Brigozzi (ed.), *Mechanisms of Speciation. Science*, 219: 1059-1060.

Gingerich, P. D. (1983) "Rates of evolution: effects of time and temporal scaling." *Science*, 222: 159-161.

Gould, S. J. (1983) "Punctuated equilibrium and the fossil record." *Science*, 219: 439-440.

Gould, S. J. and N. Eldredge (1977) "Punctuated equilibria: the tempo and mode of evolution reconsidered." *Paleobiology*, 3: 115-151.

Gregg, J. R. (1950) Taxonomy, language and reality. *American Naturalist*, 84: 419-435.

Hennig, W. (1966) *Phylogenetic Systematics*. Urbana, Ill.: University of Illinois Press.

Hull, D. L. "The effects of essentialism on taxonomy—two thousand years of stasis." *British Journal for the Philosophy of Science*, 15: 314-326; 16: 1-18.

Hull, D. L. (1974) *Philosophy of Biological Science*. Englewood Cliffs, N.Y.: Prentice Hall.

Iltis, H. H. (1983) "From teosinte to maiz: the catastrophic sexual transmutation." *Science*, 222: 886-894.

Jones, J. S. (1981) "An uncensored page of fossil history." *Nature*, 293: 427-428.

Kaufmann, E. G. (1978) "Evolutionary rates and patterns among Cretaceous Bivalvia." *Transactions of the Royal Society of London*, B 284: 277-304.

Lande, R. (1980) "Microevolution in relation to macroevolution." *Paleobiology*, 6: 235-238.

Lipscomb, D. L. (1984) "Methods of systematic analysis: the relative superiority of phylogenetic systematics." *Origins of Life*, 13: 235-248.

Malmgren, B. A., W. A. Berggren, and G. P. Lohrmann (1984) "Species formation through punctuated gradualism in planktonic foraminifera." *Science*, 224: 317-319.

Maynard Smith, J. (1982) " Evolution—sudden or gradual?" In J.

Maynard Smith (ed.), *Evolution Now*, pp. 125-128. San Francisco: W. H. Freeman.

Mayr, E. (1970) *Populations, Species, and Evolution.* Cambridge, Ma.: Belknap Press.

Mayr, E. (1981) "Biological classification: toward a synthesis of opposing methodologies." *Science*, 214: 510-516.

Mayr, E. (1982a) *The Growth of Biological Thought.* Cambridge, Ma.: Harvard University Press.

Mayr, E. (1982b) "Speciation and macroevolution." *Evolution*, 36: 1119-1122.

Newman, C. M., J. E. Cohen, and C. Kipnis (1985) "Neo-Darwinian evolution implies punctuated equilibria." *Nature* 315: 400-401.

O'Brien, S. J., D. E. Wildt, D. Goldman, C. R. Merril and M. Bush (1983) "The Cheetah is depauperate in genetic variation." *Science*, 221: 459-461.

Roughgarden, J. (1979) *Theory of Population Genetics and Evolutionary Ecology: An Introduction*. New York: Macmillan.

Simpson, G. G. (1961) *Principles of Animal Taxonomy.* New York: Columbia University Press.

Sloan, E. E., J. K. Rigby, Jr., L. M. Van Valen, and D. Gabriel (1986) "Gradual dinosaur extinction and simultaneous ungulate radiation in the Hell Creek Formation." *Science* 232: 629-633.

Stebbins, G. L. and F. J. Ayala (1981) "Is a new evolutionary synthesis necessary?" *Science*, 213: 967-971.

Williamson, P. G. (1981) "Palaeontological documentation of speciation in Cenozoic molluscs from Turkana Basin." *Nature*, 293: 437-443.

Wright, S. (1982) "The shifting balance theory and macroevolution." *Annual Review of Genetics*, 16: 1-19.

LABORATORY EXPERIMENTATION IN ECONOMICS, AND ITS RELATION TO ECONOMIC THEORY

ALVIN E. ROTH

I. Introduction

Controlled experiments conducted by economists under laboratory conditions (as distinct from field experiments[1]) have a relatively short history. Only in the last ten years has laboratory experimentation in economics completed the transition from being a seldom encountered curiosity to a well established part of the economic literature.

As an episode in the sociology of science, although it is still too early to know where it will ultimately lead, this transition has so far been rapid. For example, if I may personalize the transition with my own experience, when I first began experimental work in the early 1970's, it was most convenient to publish the results in journals of psychology and business. Today it is not unusual to find experiments reported in any of the major American economics journals. In fact, the *Journal of Economic Literature* (which is the chief abstracting journal in economics) has this year initiated a separate bibliographic category for "Experimental Economic Methods." To use a different kind of indicator, every five years the Econometric Society commissions a small number of surveys of advances made in the last five years. This year, for the first time, one of the topics surveyed was laboratory experimentation (see Roth 1985a).[2] However my topic is not the history and sociology of laboratory experimentation in economics, but rather some of the different ways in which laboratory experimentation is being employed in economics today.

Let me begin by trying to map out for you some of the different kinds of experiments that have been conducted. Perhaps the best way to do this will be by considering the relationship that various bodies of experimental work have to existing economic theories. For a start, it will be helpful to consider whether an experiment is designed to test some relatively well-defined formal theory, or to examine the effect of some manipulation about which no such theory is available. In the case of experiments that test the predictions of existing theories, it will be useful to distinguish between those that test a theory on its specified domain of application, and those that explore its applicability outside the strict confines of this domain. We will also want to know whether the results of these experiments falsify the predictions of the theory or support it.[3] Each of the experimental investigations that I will discuss consists of a set of experiments, and each overlaps in more than one of these categories (see Figure 1). There is some arbitrariness in this classification, since it often isn't clear precisely what constitutes the appropriate domain for a given theory.

To conclude my introductory remarks, let me introduce you to two bits of economic theory, the first drawn from the theory of cooperative games and the second of more ancient lineage. These will serve both to unify the discussion of much of the experimental work I am about to describe, and also give you a better idea of what I mean when I speak about the specified domain on which a theory is expected to apply.

In any economic environment, or "game," a feasible outcome is said to be *dominated* if there is some set of agents who all prefer another outcome, and who have the ability under the rules of the game to enforce that other outcome through their own (collective) actions. The *core* of the game is defined to be the (possibly empty) set of undominated outcomes. The core is thus the set of outcomes having the property that no subset of agents can arrange, by their own actions, an outcome they all prefer.

If some theory predicts an outcome outside of the core, it is predicting that some coalition of agents who all prefer another outcome,

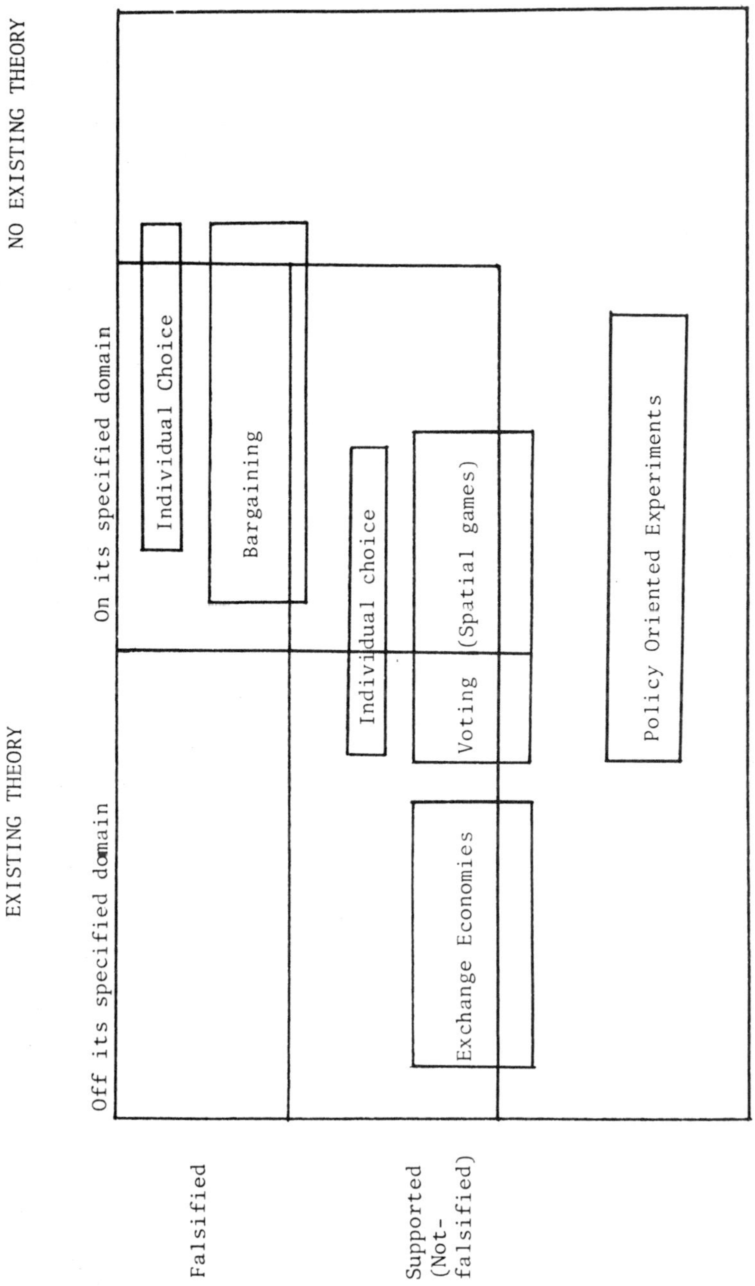
NO EXISTING THEORY
EXISTING THEORY
On its specified domain
Off its specified domain
Individual Choice
Bargaining
Individual choice
Voting (Spatial games)
Exchange Economies
Policy Oriented Experiments
Falsified
Supported (Not-falsified)

Figure 1

and who collectively have the power to produce it, will fail to do so. We might reasonably ask of such a theory why this should be the case, and it is easy to see that, in some situations, there might be very good reasons. For example, the members of the relevant coalition might not be able to identify each other (perhaps because they don't know each other's preferences), or they might not be able to trust each other to carry out the necessary steps of their collective action. For these reasons, the predictions that "outcomes will tend to be in the core" is generally regarded as having the most chance of being true in what are called cooperative games of complete information, by which is meant environments in which players are able to make binding contracts with one another and in which they know one another's preferences. For our purposes, this will be the specified domain of the "theory of the core."[4]

The second bit of theory, which will be more familiar, is the theory of competitive behavior in exchange economies, in which sellers of a good seek to transact with buyers. At any price level, the aggregate supply available is the total amount of the good that sellers are willing to offer at that price, while the aggregate demand is the total amount of the good that buyers are willing to purchase at that price. As the price rises, more units (held by more sellers) become profitable to sell, while fewer buyers find it profitable to make purchases, so the aggregate supply rises and the aggregate demand falls. A competitive equilibrium occurs when all units are transacted at the price for which aggregate supply equals aggregate demand. (At this price, the market clears: the amount offered for sale equals the amount that buyers wish to purchase.) The prediction that markets will reach competitive equilibrium is therefore a prediction about both the quantity of goods that will be sold, and their price.

The most commonly invoked formal models of competitive behavior are specified for the idealized situtation of "perfect competition," in which infinitely many agents of negligible size act as "price takers" who passively respond to market prices as individual optimizers.[5] This is a domain that can obviously not be produced in the laboratory. However the competitive equilibrium can be computed whether or not these conditions are met, and we will see that its predictions have received substantial support in laboratory markets involving relatively small numbers of buyers and sellers all of whom are actively involved in setting prices, under certain rules of market organization.

II. Voting in spatial games

Spatial games involve some group decision process to determine the location of a public facility, where the preferences of the agents are related to their own, fixed, locations. In the two experiments I will discuss here,[6] which were both designed to test the predictions of the core, subjects were instructed that they would collectively determine, by a sequence of proposals and votes that would be decided by majority rule, a point on the plane. The subjects were all informed that each would receive a monetary payoff whose amount depended on how close the point decided on came to that subjects's own "ideal point," and each subject was informed of the coordinates of his own ideal point. The most important difference between the two experiments was that the first was conducted under conditions of complete information: each agent was informed of the other's (ordinal) preferences by being informed of the location of all of the ideal points. In the second experiment, each agent was informed only of the location of his own ideal point, and so did not know the preferences of the other agents. Thus the first experiment is a test of the theory of the core on it specified domain, and the second is a test of the theory's predictions outside of this domain.

In both experiments, the particular games studied provided a strong test of the core in the sense that the core consisted of a unique outcome, rather than a large set of outcomes. However there is another sense in which these experiments provided only a weak test of the core as such: in many of the games studied, the unique point in the core was not merely the unique undominated outcome, but also an outcome that itself dominated all others. Consequently a number of different game theoretic theories that depend on the domination relation all yield the same prediction in these games. For these games, then, observing the core outcome lends support to theories based on the domination relation, rather than specific support to the core.

The outcomes observed in both experiments were often different from the (unique) core outcome, but tended to cluster around it. A particularly effective graphic device is used by Fiorina and Plott to display their initial results, which are not atypical of the other results in these and related experiments. They considered the (not always distinct) outcomes predicted by sixteen different theories drawn from the literature of economics, political science, psychology, and sociology, and displayed their data on a grid that divided the plane into the regions closest to each predicted outcome. (Thus a data point that appears in the region containing the core represents an observed outcome that is closer to the core than to any other of the predicted outcomes.) They found that the core was the modal observed outcome, that approximately 75% of the observed outcomes fell in the region closest to the core, and that the remaining outcomes fell in regions adjacent to the core. These experiments therefore support the hypothesis that the core has predictive power. This predictive power was robust with respect to whether or not subjects knew each other's preferences, which suggests tht repeated voting serves to effectively disseminate some of this information.[7]

III. Bargaining

The experiments I will discuss next, in which I have personally been involved, are concerned with two-person bargaining under the following rules. The two agents may allocate some valuable resource between themselves in any way they like, provided they both agree. If they fail to agree on how to allocate the resource, they each receive nothing. This is an example of what is sometimes called a "pure" bargaining game, which can be thought of as the opposite of the idealized case of "perfect" competition: whereas individual agents have negligible influence under perfect competition, in a pure bargaining game each agent has an absolute veto over every division of the resource between the bargainers.

The core of such a game is the large set of outcomes corresponding to the entire set of agreements that leave no part of the resource unallocated. The prediction that observed outcomes will be in the core is therefore a fairly weak (although not an empty) prediction. There has thus been considerable sustained interest in developing theories that attempt to predict specific outcomes in the core. Such theories attempt to make use of information concerning some measure of the *intensity* of agents' preferences, and most often do so by using the kind of information contained in an expected utility function representing those preferences.[8] That is, these theories attempt to predict the outcome of bargaining on the basis of information about each bargainer's willingness to tolerate risk.

Because an agent's willingness to tolerate risk cannot be observed in uncontrolled, naturally occuring, environments, theories of this kind proved unusually difficult to test with field data. Furthermore, because the theories made their clearest predictions under conditions of "complete information" in which bargainers were assumed to know one another's expected utility functions, the requirements of the theory were sometimes treated as an idealization incapable of a concrete realization.

To provide a test of the predictions of these theories in the laboratory, an experimental design was developed (see Roth 1979 and Roth and Malouf 1979) that controlled for the risk preferences of the bargainers. Risk neutral preferences were induced by having the experimental subjects bargain (anonymously, via computer terminals) over the distribution of "lottery tickets" that determined the probability that each bargainer would receive a certain fixed cash prize (possibly a different prize for each bargainer) or receive nothing. The lottery tickets were the valuable resource to be divided: if the bargainers failed to reach agreement, they each received nothing (and hence no chance to win their cash prize), while if they reached an agreement giving, say, 80% of the lottery tickets to one of the bargainers, then that bargainer had an 80% chance of winning his prize and a 20% chance of winning nothing. Since the information about preferences conveyed by an expected utility function is meaningfully represented only up to the arbitrary choice of origin and scale, there is no loss of generality in normalizing each agent's utility so that his utility for receiving his prize equals 1, and his utility for receiving nothing equals 0. The utility of each agent for any agreement is then precisely equal to his probability of receiving his prize, i.e. equal to the percentage of lottery tickets he receives.[9] Consequently bargaining under this design meets the requirement of being a game of "complete information," since the information determining each agent's utility for any agreement is common knowledge. (Note that this is so independently of the value of the monetary prizes, and whether each bargainer knows the other's prize.) The design thus provides the opportunity to test various economic theories of bargaining precisely on the domain to which they are intended to apply, and to examine what if any effect might be due to information about the prizes in addition to the "complete" information of the formal models.

A series of experiments using this basic design[10] found that information about the relative size of the bargainers' cash prizes influenced the outcome of bargaining in a way inconsistent with the predictions of existing economic theories of bargaining. When bargainers did not know the value of each other's prize, agreements clustered around the 50-50 division of lottery tickets. However when bargainers knew the value of each other's prize, the bargainer with the smaller prize tended to receive a higher percentage of the lottery tickets. This did not occur "smoothly": instead, a bimodal distribution of agreements was observed, together with an increase in the percentage of negotiations that ended in disagreement. Furthermore, this bimodal distribution of agreements produced by information about the cash value of prizes was *not* reproduced by "strategically equivalent" information (about artificial commodities, that does not allow actual cash values to be compared) that existing theory predicts should be equivalent.

The frequency of disagreement turns out to depend in subtle ways on how information is distributed. For example, it makes a difference whether information about prizes is common knowledge: e.g. it matters not only whether bargainers know each other's prizes, but also if they *know* whether each knows the other's prize. However the data show that the frequency of disagreement exhibits a pattern characteristic of economic equilibrium, in that there is a precise sort of tradeoff between how much a bargainer demands and his probability of reaching agreement. The data show that the rate at which the probability of reaching agreement declines as final demands increase is very close to the rate at which a bargainer should be indifferent between changing his final demand or sticking to it. This strongly suggests that the bargainers are evaluating their prospects correctly, and behaving neither too aggressively nor too timidly.

There are other observable regularities of bargaining behavior.

Some of these were observed in experiments conducted on a slightly more complex domain than binary lottery games, in order to allow scope for the expression of differences among bargainers in their willingness to take risks.[11] The observed behavior supports some of the most subtle and general qualitative predictions of the existing theories about the influence on bargaining success of a bargainer's willingness to tolerate risk.[12] The regularities in the observed outcomes thus suggest that these phenomena will be amenable to economic analysis. Another experiment[13] lends support to the hypothesis that the (unpredicted) effect of information about the cash value of prizes is attributable to its effect on the expectations of the bargainers about what constitutes a credible bargaining position. I am inclined to think that this will be a fruitful direction in which to explore new theories of bargaining able to accomodate the kind of behavior observed in these experiments.

IV. Exchange economies

The experiments I will talk about next involve markets in which buyers and sellers seek to trade with one another under one of a variety of rules of market organization.[14] In each market, sellers are told how many units of a good they may "produce," and given a schedule of unit costs—i.e. a seller might be told that he may sell up to two units, the first of which will cost him $1.00 to produce and the second $1.25. At the conclusion of the experiment, each seller is paid the sale price of any unit he has sold minus its cost: i.e. if the above seller sells only one unit, at a price of $1.10, he will receive (a profit of) $0.10 for the transaction. Similarly buyers are told how many units they can consume, and what the value to them of each unit is. (All goods are indivisible and homogeneous—they can only be consumed in whole units, and one unit is identical to another, regardless of who the seller is.) That is, a buyer might be told he can only consume one unit, and it is worth $1.00 to him. At the conclusion of the experiment, each buyer is paid the value to him of any units he has purchased, minus the purchase price. Thus if the above buyer makes a purchase at a price of $0.75, he will receive $0.25 for the transaction.[15] Each agent knows only his own costs or values.

Various rules for making transactions were compared in these experiments. In a *double oral auction* sellers may make offers and buyers may make bids. Sellers and buyers may make new offers and bids whenever they like; the lowest outstanding offer is the "market asked price," and the highest outstanding bid is the "market bid price." If any seller offers the market bid price, or if any buyer bids the market asked price, a transaction is consummated between the seller and buyer whose prices coincide.

In a *single oral auction* only one side of the market is allowed to suggest prices. There are thus two kinds of single oral auctions, the *bid* oral auction in which buyers may make bids, and the *offer* oral auction in which sellers may make offers. Agents on the passive side of the market may merely accept the current market (bid or asked) price, or remain silent. Agents on the active side of the market may revise their prices whenever they wish. A transaction occurs when an agent on the passive side of the market accepts the market price.

In a *posted price market,* as in the single oral auction, only one side is allowed to suggest prices. However each agent on the active side of the market must "post" his price when the market opens, and is thereafter not free to change it. There are of course two kinds of posted price markets, depending on whether sellers are posting offers or buyers are posting bids.

In each of these experiments, some small numbers of buyers and sellers (typically from 4 to 12 of each) engaged in some moderate number of repetitions of such a market (typically 5 to 15 "trading days") under one of the market institutions described above. Each trading day constitutes a new market, with the same agents and parameters. (Thus in the various oral auctions agents could revise their prices as often as they wished each trading day, but in the posted price markets, prices could be posted only at the beginning of each trading day.) The final payment that subjects received was equal to their total profit on all transactions in all trading days.

In all of the experiments, transaction prices in the final period were observed to be significantly closer to the competitive price than in the first period. This was most clearly observed in the double oral auction markets, in which many transactions in the final periods occurred precisely at the competitive price, and in which the quantity traded was virtually always within a single unit of the competitive quantity.

The prediction of the competitive model thus receives support from the double auction data. This is somewhat surprising in view of the small numbers of traders involved, and the active role that each plays in determining the price. This is clearly not an environment that approximates "perfect" competition as it has traditionally been formulated. What seems to be happening is that a great deal of information is conveyed to the agents in this kind of market, not merely by the transactions that take place, but also by the bids and offers that are not accepted and thus *don't* lead to transactions. For example, if the first few trades that take place on the first trading day happen to occur at a price far above the competitive price, then some potential buyers will be priced out of the market. As the trading day progresses, sellers who have units left to sell lower their asking prices a little, hoping to transact with some of these buyers. All the buyers learn from these lower asking prices that some units are available at a lower price than previous transactions, and this influences the bids made on the next trading day. Extending the domain of the theory of competitive behavior to small double auction markets will likely involve paying attention to how this kind of information tells buyers and sellers "what the market will bear."

While the other market institutions also exhibited some tendency to converge towards the competitive price and quantity, as compared with the results of the double oral auction markets they each exhibited a bias toward one or the other side of the market. Curiously, the bias in the single oral auctions is in favor of the passive side of the market, while the bias in the posted price markets is in favor of the active side. That is, compared to the prices observed in the oral double auction (and to the competitive price), a single oral auction yields higher prices when the buyers can make bids, and a lower price when the sellers can make offers, while the posted price market yields a higher price when the sellers can post offers, and a lower price when the buyers can post bids. Thus it is an advantage to be on the active side of a posted price market rather than on the passive side, and an advantage to be on the passive side of a single oral auction rather than on the active side. No presently available theory that I know of is able to account for these results.

V. Policy oriented experiments

The next experiment I will discuss[16] falls squarely in the category of experiments that test the effect of changes for which no formal theory is available. This type of experiment is representative of those that are motivated by questions of policy of the kind raised by government regulatory agencies, typically about the effect of changes in the rules by which some market is organized. These investigations offer the possibility of bringing scientific methods to bear on one of the traditional nonscientific vocations of economists, which is whispering in the ears of princes who require advice about pressing practical questions whose answers lie beyond the reliable scientific knowledge of the profession.

The study I will speak about arose in a matter of concern to the Interstate Commerce Commission, and had to do with a complex posted price market. The case concerned whether barge operators should be required by the ICC to post their prices, and announce price changes in advance. The existing market allowed rates to be set by individual negotiations between barge operators and their customers, so that the terms of each contract were private information. Plott (forthcoming) reports that the question arose because railroad companies were lobbying to require such price posting. The reasons offered by the railroads were that

> . . . the public information feature of posted rates would make the industry more competitive, allow the railroads to compete better, and aid the small barge owners who were allegedly secretly being undersold by the large barge companies. [Plott, forthcoming]

In their introductory comments, Hong and Plott (1982) say the following about their use of laboratory experimentation to illuminate the issues raised by the proposed change.

> The full consequences of a rate filing policy are unknown. Plausible theoretical arguments can be made on both sides of the policy argument. When existing theory does not yield a definitive answer, one can usually turn to previous experiences

> with policies, but in this case we are aware of no industrial case study that would provide direct evidence on either side of the controversy.

They go on to note that it would be difficult to draw any compelling policy conclusions regarding the barge industry from previous laboratory experiments concerning posted price markets, since

> Any extrapolation from published experimental results to the barge industry itself is open to two potential criticisms, the reasonableness of which this study was designed to assess. First, the barge industry has several prominent economic features that are not incorporated in existing laboratory market studies. Examples include the relative sizes of buyers and sellers, the demand and supply elasticities, and the cyclical nature of demand. Naturally, we can never be certain that all the important features have been included in the present design. If something important has been misspecified or omitted, then the observed behavior of the laboratory market may not extend to the barge industry, and additional appropriately modified experiments can be conducted as checks on our conclusions. The second potential criticism is that the effects of price posting in laboratory studies have only been measured relative to the performance of oral auction markets. Since auction markets differ from the negotiated price markets of the industry, the relevance of the comparison can be questioned.

Hong and Plott proceeded to design their experiment around a laboratory market scaled to resemble, in the features mentioned above, the market for transporting grain along the upper Mississippi River and Illinois Waterway during the Fall of 1970. (This market was chosen because it was believed to be representative of a significant portion of the dry bulk barge traffic in the United States, and because adequate data about the market parameters were available.) Aggregate supply and demand functions for the laboratory market were scaled to estimates available for the target market, as was the distribution of large and small firms on each side of the market. The laboratory market was divided into periods representing two weeks of the target market, and the seasonal aspects of the target market were modelled by having demand in the laboratory market scaled to resemble two months of normal demand followed by two months of high demand, followed by two months of normal demand. The experimental design involved running the market under both posted price and negotiated price policies.

In presenting the data from this experiment, Hong and Plott report that

> The results are easy to summarize. The posted price policy causes higher prices, reduced volume, and efficiency losses. Furthermore, the posted price policy works to the disadvantage of most market participants, especially the small ones, and helps only the large sellers.

They also conclude that the posted price markets react more slowly to the seasonal change in demand than do the negotiated price markets.

Plott (forthcoming) reports that this experimental evidence helped to deter the lobbying of the railroads on this matter, and that the price posting policy that they had advocated was not pursued. He aptly characterizes the role of laboratory experimentation in this kind of policy debate as serving primarily to *shift the burden of proof*. This may not in fact be completely different from the role that experiments play in testing hypotheses derived from formal theories. But there are some clear differences between designing an experimental market to resemble some complex naturally ocurring environment, in the hope of making observations that may be relevant to a particular issue of policy, and designing an experiment to test some particular formal hypothesis, in the hope of making observations that may support a fairly general conclusion.

VI. Some important omissions, and concluding remarks

There are of course many other experiments I could have discussed. This paper is intended as a sampling rather than as a comprehensive survey of experimental economics. The variety of experimental studies that economists have undertaken would make such a survey a major undertaking, and the rapid appearance of new work would make it quickly obsolete. There are, however, two bodies of work that I did not discuss, but which I want to mention at least briefly, since they may be of particular interest to philosophers of science. Both concern theories of individual choice.

The first of these is the body of work concerned with experimentally examining people's *individual* choice behavior (as opposed to their collective behavior in economic environments such as markets). This work has been particularly concerned with testing the implications of the theory that people are Bayesian decision makers, whose choice behavior can be modelled via von Neumann Morgenstern utility functions. Since this theory is at the foundation of much of economic theory, this is a subject that has aroused considerable interest among economists. I am thinking here not merely of the work of economists such as Maurice Allais (1953), who observed some systematic violations of expected utility theory, but also of the well known recent work of psychologists such as Daniel Kahneman and Amos Tversky.[17] This latter work explores not only observed violations of Bayesian decision theory, but also regularities in the way individual choice behavior responds to variables about which there exists little in the way of (prior) theoretical predictions.

This work has elicited a variety of responses from economists. Some (see e.g. Thaler, forthcoming) believe that these results call for a fundamental reexamination of a large part of economic theory. Others (see e.g. Knez and Smith, forthcoming) present experimental evidence which they interpret as suggesting that markets may act to "smooth out" some of the unpredicted individual choice behavior, so that much of the experimental evidence about the shortcomings of expected utility theory may not be of great significance for economic theory. There has also been renewed effort to construct alternatives to expected utility theory, that are capable of explaining the experimental data on individual choice behavior. Some of the results in this direction (see e.g. Machina, 1982) suggest that many of the features of expected utility theory that economists make the most use of (such as predictions about risk posture) will survive the transition to theories that are more descriptive of individual choice behavior in the experimental environments that have been examined.

The other body of work that I want to mention here concerns experiments that have been conducted to test the applicability of simple economic theories of individual choice—specifically, theories of consumption and labor supply—to the behavior of laboratory animals, namely pigeons and rats. (See particularly Kagel, forthcoming; and Kagel, Battalio, and Green, 1981). In these experiments, the laboratory animals are observed in controlled environments where they may perform specific tasks (such as pecking a key for pigeons, or pressing a lever for rats) to obtain food or water. By changing the number of key pecks required to obtain a given amount of food or a given amount of water, the relative prices of food and water can be changed, and observations about changes in the consumption of food and water (and leisure) can be compared to the prediction of economic theories. There is a large body of theory and experimental evidence on animal behavior in the behavioral psychology literature, and it appears that economic theories of behavior compete quite effectively in terms of their predictive and explanatory power.

6.1 Concluding Remarks

It should be clear that there are many ways to classify economic experiments other than the one I have chosen here.[18] In Roth (1985a) I tried to classify experiments by what kind of dialog they were a part of, with some experiments being primarily part of a dialog between experimenters and theorists, some primarily part of a dialog between experimenters and other experimenters, and some primarily part of a dialog between experimenters and policy makers.[19] There has not yet developed a substantial dialog between experimenters and empirical economists who work with field data, but it is clear that this is one of the directions in which there is great potential.

The role of experimental methods in economic research is continuing to evolve, and it is too early to know all the ways in which experimentation can interact with more traditional tools of economic research. The ultimate place of experimentation in the economics profession's portfolio of tools will likely be different than the place of experimentation in other sciences. Still, there are some analogies that may be helpful. Like economics, biology and astronomy are both concerned in large part with historical phenomena that cannot be brought into the laboratory. The evolution of species and of galaxies can no more be studied experimentally than can large scale macroeconomic phenomena. But biologists' understanding of evolution is enhanced by experiments in plant breeding, genetics, and molecular biology, and astronomers' understanding of cosmology is enhanced by experiments in mechanics and particle physics. In the same way, microeconomic experiments, through the impact they have on microeconomic theory, will also have an impact on economists' understanding of macroeconomic phenomena. As in other sciences, the role of laboratory experimentation in economics rests on the fact that it gives us a way to answer certain kinds of questions that we have no other way of answering.[20]

NOTES

1. See e.g. Ferber and Hirsch, 1982, Or Hausman and Wise, 1985.

2. Some of the descriptive material in this paper is borrowed from my lengthier discussion there.

3. I know that, for this audience, I should probably not say that evidence ever "supports" a theory: please fell free to understand me as saying that the evidence fails to falsify the theory.

4. Economists often consider other ways to restrict this domain or soften the prediction, but for simplicity and brevity I will not discuss these.

5. For example, it is straightforward to show that, for any exchange economy, the "competitive allocation," which is the outcome that results when a competitive equilibrum is reached, is always contained in the core. However when the market consists of infinitely many agents of negligible size, the core contains *only* competitive allocations.

6. Berl, Mckelvey, Ordeshook, and Winer, 1976; and Fiorina and Plott, 1978

7. Interestingly, subjects who were debriefed after participating in these games often expressed the opinion that the final outcome was determined largely by chance.

8. The most influential single theory of this sort is one due to John Nash(1950) whose work led to the development of a family of related theories in the tradition of cooperative game theory (see Roth, 1979 for a full account). More recently, there has been a growing body of work on theories of bargaining in the tradition of noncooperative game theory (see Roth, 1985c for a collection of representative work).

9. Note that restricting the agents to an environment in which all choices involve lotteries over only two ultimate payoffs (receiving your prize or receiving nothing) controls out the effects of risk aversion or risk preference, which are phenomena that arise in making tradeoffs among three or more outcomes. Also, the regularities needed for an agent's preferences to be representable by an expected utility function on this domain reduce to the simple requirement that a higher probability of winning his prize be preferable to a smaller probability. So some of the individual choice phenomena inconsistent with a utility representation that have been observed on more complex choice domains are also controlled out in this design by restricting the domain of choices in this way.

10. Roth and Malouf (1979); Roth, Malouf, and Murnighan (1981); Roth and Murnighan (1982); see also Roth (1983).

11. See Murnighan, Roth, and Schoumaker (1985).

12. Risk posture is obviously an important factor in the predictions of theories that are stated in terms of the von Neumann-Morgenstern utilities of the bargainers. It turned out, however, that the predictions of these theories about the specific effects of risk aversion had not been developed in a way that yielded experimentally testable hypothesis. (One indirect virtue of experimentation is that it can provide a discipline to theoretical work, and suggest directions in which theory ought to be explored.) To this end, therefore, a systematic study of the predictions of these models relating to the risk posture of the bargainers was carried out in Roth (1979), Kihlstrom, Roth, and Schmeidler (1981), Roth and Rothblum (1982), and Roth (1985b). Rather surprisingly, a very broad class of apparently quite different models yield a common prediction regarding risk aversion. Loosely speaking, these models all predict that risk aversion is disadvantageous in bargaining, except when the bargaining concerns potential agreements that have a positive probability of yielding an outcome that is worse than disagreement.

13. See Roth and Schoumaker (1983).

14. See Smith (1964), (1976), Plott and Smith (1978), and Plott (1982).

15. In the experiments discussed here, agents were effectively forbidden to make trades that would bring them a negative return.

16. Hong and Plott (1982).

17. See e.g. Kahneman and Tversky (1979) or Tversky and Kahneman (1981).

18. See Roth (forthcoming) for a collection of papers exemplifying the different points of view of a number of influential economic experimenters.

19. Note that I haven't tried to *define* what an economic experiment is. Even if such an endeavor might eventually be possible and useful, the range of laboratory investigations that economists will find it useful to en-

gage in has yet to be explored enough to try to draw boundaries. (For example, much of the material covered here falls outside the boundaries of Smith's (1982) definition of what constitutes an economic experiment.)

20. This work has been supported by grants from the National Science Foundation and the Office of Naval Research, and by a Fellowship from the Alfred P. Sloan Foundation.

REFERENCES

1. Allais, Maurice, "Le comportement de l'homme rationnel devant le risque: critique des postulats et axiomes de l'ecole americaine," *Econometrica* vol. 21 (1953), pp. 503-546.

2. Berl, Janet E., Richard D. McKelvey, Peter C. Ordeshook, and Mark D. Winer, "An Experimental Test of the Core in a Simple N-Person Cooperative Nonsidepayment Game," *Journal of Conflict Resolution*, vol. 20 (1976), pp. 453-479.

3. Ferber, Robert, and Werner Z. Hirsch, *Social Experimentation and Economic Policy* (Cambridge: Cambridge University Press, 1982).

4. Fiorina, Morris P. and Charles R. Plott, "Committee Decisions under Majority Rule: An Experimental Study," *The American Political Science Review*, vol. 72 (1978), pp. 575-598.

5. Hausmann, Jerry A. and David A. Wise (eds.) *Social experimentation* (Washington, D.C.: National Bureau of Economic Research, 1985).

6. Hong, James T. and Charles R. Plott, "Rate filing policies for inland water transportation: an experimental approach," *Bell Journal of Economics* vol. 13 (1982), pp. 1-19.

7. Kagel, John H., "Economics According to the Rats (and Pigeons Too): What Have We Learned, and What Can We Hope to Learn," in Alvin E. Roth (ed.), *Laboratory Experimentation in Economics: Six Points of View* (Cambridge: Cambridge University Press, forthcoming 1987).

8. Kagel, John H. and Raymond C. Battalio, Leonard Green, "Income-Leisure Tradeoffs of Animal Workers," *American Economic Review*, vol. 71 (1981) pp. 621-632.

9. Kahneman, Daniel and Amos Tversky, "Prospect Theory: An Analysis of Decision under Risk," *Econometrica*, vol. 47 (1979) pp. 263-291.

10. Kihlstrom, Richard, Alvin E. Roth and David Schmeidler, "Risk Aversion and Solutions to Nash's Bargaining Problem," in O. Moeschlin and D. Pallaschke (eds.), *Game Theory and Mathematical Economics* (Amsterdam: North-Holland, 1981), pp. 65-71.

11. Knez, Mark and Vernon L. Smith, "Hypothetical Valuations and Preference Reversals in the Context of Asset Trading," in Alvin E. Roth

(ed.), *Laboratory Experimentation in Economics: Six Points of View,* Cambridge: Cambridge University Press, forthcoming 1987).

12. Machina, Mark J., 'Expected Utility' Analysis Without the Independence Axiom," *Econometrica,* vol. 50 (1982), pp. 277-323.

13. Murnighan, J. Keith, Alvin E. Roth, and Francoise Schoumaker, "Risk Aversion in Bargaining: An Experimental Study," working paper, 1985.

14. Nash, John, "The Bargaining Problem," *Econometrica,* vol. 28 (1950), pp. 155-162.

15. Plott, Charles R., "Industrial Organization Theory and Experimental Economics," *Journal of Economic Literature,* vol. 20 (1982), pp. 1485-1527.

16. Plott, Charles R., "Dimensions of Parallelism: Some Policy Applications of Experimental Methods," in Alvin E. Roth (ed.) *Laboratory Experimentation in Economics: Six Points of View* (Cambridge: Cambridge University Press, forthcoming 1987).

17. Plott, Charles R. and Vernon L. Smith, "An Experimental Examination of Two Exchange Institutions," *Review of Economic Studies,* vol. 45 (1978), pp. 133-153.

18. Roth, Alvin E. *Axiomatic Models of Bargaining* (Berlin: Springer Verlag, 1979).

19. Roth, Alvin E. "Toward a Theory of Bargaining: An Experimental Study in Economics," *Science,* vol. 220 (May 13, 1983), 220, pp. 687-91.

20. Roth, Alvin E. "Laboratory Experimentation in Economics," in Truman Bewley (ed.), *Advances in Economic Theory, 1985,* (Symposia of the 5th World Congress of the Econometric Society), (Cambridge: Cambridge University Press, 1985a, forthcoming).

21. Roth, Alvin, E., "A Note on Risk Aversion in a Perfect Equilibrium Model of Bargaining," *Econometrica,* vol. 53 (1985b), pp. 207-211.

22 Roth, Alvin E. (ed.), *Game-Theoretic Models of Bargaining* (Cambridge: Cambridge University Press, 1985c).

23. Roth, Alvin E. (ed.) *Laboratory Experimentation in Economics* (Cambridge: Cambridge University Press, forthcoming 1987).

24. Roth, Alvin E. and Michael W. K. Malouf, "Game-Theoretic Models and the Role of Information in Bargaining," *Psychological Review,* vol. 86 (1979), pp. 574-594.

25. Roth, Alvin E., Michael W. K. Malouf, and J. Keith Murnighan, "Sociological Versus Strategic Factors in Bargaining," *Journal of Economic Behavior and Organization,* vol. 2 (1981), pp. 153-177.

26. Roth, Alvin E. and J. Keith Murnighan, "The Role of Information in Bargaining: An Experimental Study," *Econometrica,* vol. 50 (1982), pp. 1123–1142.

27. Roth, Alvin E. and Uriel G. Rothblum, "Risk Aversion and Nash's Solution for Bargaining Games With Risky Outcomes," *Econometrica,* vol. 50 (1982), pp. 639-647.

28. Roth, Alvin E. and Francoise Schoumaker, "Expectations and Reputations in Bargaining: An Experimental Study," *American Economic Review,* vol. 73 (1983), pp. 362-372.

29. Smith, Vernon L., "Bidding and auctioning institutions: experimental results," in Y. Amihud (ed.), *Bidding and auctioning for procurement and allocation,* (New York: New York University Press, 1976).

30. Smith, Vernon L., "Effect of Market Organization on Competitive Equilibrium," *Quarterly Journal of Economics,* vol. 78 (1964), pp. 181-201.

31. Smith, Vernon L., "Microeconomic Systems as an Experimental Science," *American Economic Review,* vol. 72 (1982), pp. 923-955.

32. Thaler, Richard, "The Psychology of Choice and the Assumptions of Economics," in Alvin E. Roth, (ed), *Laboratory Experimentation in Economics: Six Points of View,* (Cambridge: Cambridge University Press, forthcoming 1987).

33. Tversky, Amos and Daniel Kahneman, "The framing of decisions and the psychology of choice," *Science,* vol. 211 (1981), pp. 453-458.

WHITHER COMPLEMENTARITY?

MICHAEL REDHEAD

I. Introduction

COMPLEMENTARITY was introduced by Bohr in his famous Como lecture in 1927. Versions were subsequently published in Nature and in his book *Atomic Physics and the Description of Nature*. Let me begin by sketching very briefly the experiments which led to the doctrine of complementarity. These provided a *prima facie* case for wave-particle duality for material particles such as electrons and also for radiation. Historically radiation came first.

In 1905 Einstein proposed the photon concept, that electromagnetic radiation exhibited a particle aspect. This was based on a discussion of the thermodynamics of the radiation field, but was also applied to the photo-electric effect in which the energy of the emitted electron was controlled by the frequency and not the intensity of the incident radiation. The particle aspect of light was given further spectacular support by the discovery of the Compton effect, in which laws of conservation of energy and momentum applied to the impact of photons and electrons explained the change in frequency of the scattered radiation via the $E = h\nu$ equivalence between energy E and frequency ν of the photon.

So much for the particle aspect of radiation. The wave aspect was demonstrated in familiar interference and diffraction phenomena.

Turning to matter rather than radiation, the particle aspect of electrons was shown by the deflection of electron beams in electrostatic and magnetic fields. These were the classic experiments of J.J.

Thomson and others. The wave aspect appeared in electron diffraction phenomena first demonstrated by Davisson and Germer (1927) and confirmed by G.P. Thomson (1928).

So the basic puzzle reads: How can something behave sometimes like a particle and sometimes like a wave, i.e. apparently exhibiting *contradictory* natures?

The simple nub of Bohr's resolution is to say that the experiments in which the particle aspect is displayed are not the same experiments as those in which the wave aspect is displayed. There is no *single* experiment in which both contradictory aspects are displayed. So no logical contradiction arises if the nature of the electron or the photon is relativized to specified experimental contexts. Wave phenomena and particle phenomena are *mutually exclusive* and yet both must be taken account of if a *complete* account is to be given of the nature of matter or radiation. Phenomena related in this way are said by Bohr to be *complementary*.

What I have given is just a popular sketch of the doctrine. In a moment I will be more careful in trying to expound what Bohr was getting at. But first let us keep to the broad outline.

What does it say about Reality?

Well it contrasts with the following classical realist theses:

(A) The world is comprised of objects, be they particles or fields, which have a fixed nature.
(B) We can discover what this nature is by "observing" these objects in the sense not necessarily of direct sense experience but of subjecting them to experimental investigation.
(C) The nature of the objects so revealed does not change according to the type of experiment carried out.

Now complementarity could be taken in two senses:

(1) The objects have definite natures but these can be changed or "disturbed" by doing experiments. Crudely a particle can be turned into a wave and vice-versa.
(2) The objects do not have definite natures at all. All that experiments reveal are appearances—we cannot infer

> backwards from the appearance to any thing-in-itself that, ontologically speaking, can be related in its nature to these "contradictory" appearances. The best we can do is to "paste" together all the appearances and this provides the most complete *description* we can give of the physical object. Any attempt to impute the appearances directly to the object would lead to a contradictory state of affairs in the physical world.

The disturbance approach (1) is definitely not to be attributed to Bohr, despite some loose formulations in his earlier writings (this particularly vitiates the arguments in the famous Bohr-Rosenfeld paper on the measurability of fields[1]). As Bohr warns us:

> One sometimes speaks of 'disturbance of phenomena by observation' or 'creation of physical attributes to atomic objects by measurements'. Such phrases, however, are apt to cause confusion, since words like phenomena and observation are here used in a way incompatible with common language and practical definition. (Bohr (1958) p. 73.)

With regard to (2) this is likely to lead to an instrumentalist gloss. As Bohr once remarked: "There is no quantum world" (Peterson (1963) p. 12). But this is an isolated remark. The overall pattern of his thinking seems to be that there is an objective world giving rise to our experience (i.e. the results of experiments), but that a limitation arises in describing such a world due to the necessity of using *classical concepts* in the description. For a convincing defence of Bohrean realism see Folse (1985). Of course the realism is not that of classical physics, outlined above. But the holistic involvement of micro-objects and experimental arrangements in exploring the appearances of the former is not to be taken to mean that they literally do not exist!

With this preamble I will consider some of the themes in Bohr's writings in more specific detail.

II. What is a Particle and What is a Wave?

In classical physics a particle has two important attributes:

(1) It is a localized entity, perhaps even a mathematical point, which moves in a well-defined trajectory through space, the motion being continuous with respect to the passage of time.
(2) It carries energy and momentum which are also localized according to the location of the particle.

A wave is typically spread out through a region of space. Even if it starts as a local concentration—a wave-packet—dispersion may lead to a spreading of the wave-packet as time proceeds. A wave also carries energy and momentum, but these are also spread through the same extended spatial region as the wave. A wave propagates by integrating the superimposed disturbances arising from all parts of its previous location. This mode of propagation leads to characteristic diffraction and interference phenomena.

What does Bohr mean by saying that an object exhibits wave-like behaviour? This is always in reference to diffraction and interference effects. These arise typically in two ways:

(1) Partial localization of the object by passing it through a single slit that is not *too* narrow.
(2) Producing an ambiguity in localization by passing it through a double slit arrangement, in which the slits can be narrow as we please.

We can thus identify the wave aspect as generated by *partial* localisation. But it is the apparent mode of propagation which is really at issue here.

Turning to particle-like behaviour, Bohr seems always to have in mind two cases:

(1) The way the object exchanges energy and momentum with the observing apparatus. There is a discreteness in-

volved here, that the wave picture cannot easily accommodate, for example a beam of electrons when absorbed by a screen communicates to the screen the discrete sum of the momenta carried by each electron.

(2) Localization, as for example in observing through which slit in the 2-slit experiment the electron passes.

The conflict between wave and particle aspects is often seen as a *logical* contradiction. But our analysis has shown that that is not right.

The mode of propagation does not show that we *cannot* be dealing with particles, subject to appropriate "funny" forces. Indeed a detailed presentation of how this can be done has been given in the work of Bohm and his colleagues.[2] Similarly the discreteness associated with the particle picture does not contradict a wave or at any rate a field ontology. Redhead, for example, has stressed in his (1983) the way in which classical particle physics can be interpreted as a field theory.

Many interpreters of Bohr, particularly Rosenfeld (cf. his (1953), have claimed that the complementarity resolution of wave-particle duality is *forced* on us due to the wave and particle aspects being *logically* irreconcilable. Thus, says Rosenfeld (p. 394):

> The conception of complementarity forces itself upon us with logical necessity. It is not some fanciful speculation which we could at will accept or reject according to whether we find it conformable to some philosophical criteria or other . . . We [are] led to modify the theory of knowledge; we cannot possibly change the lessons of experience.

In fact the metaphysical position here is quite *underdetermined* by the phenomena. That is not to say that arguments may not be adduced for and against complementarity, as we shall discuss presently, but the arguments are not simply ones of empirical adequacy, what is required to *save the phenomena*.

III. Space-time Coordination and the Claim of Causality

Although Bohr was wont to preface his papers on complementarity with reference to wave-particle duality, he always went on to stress the complementarity between *space-time coordination* and what, in his Como lecture, he refers to as the *claim of causality*. By space-time coordination Bohr clearly means the localization in space and time of micro-entities. By the claim of causality he understands the appropriateness of assigning well-defined momentum and energy to micro-entities, since causality is identified by Bohr with the operation of conservation laws for these quantities.

It has often been pointed out (see e.g. Scheibe (1973) p. 35) that causality in classical physics, in the sense of the operation of a causal process, involves a *combination* of space-time coordination and the dynamical conservation laws. But this is just an instance of the way Bohr uses familiar terminology in a somewhat special and idiosyncratic way.

The complementarity of space-time coordination and the claim of causality is clearly linked to the familiar uncertainty relations affecting x and p or E and t where x is a position coordinate and t a time coordinate for a particle with momentum p and energy E. For Bohr the limitations on joint *definability* of these conjugate attributes does not arise as a result of the uncertainty relations if these are understood as a limitation on possible knowledge. It is rather the other way about. The complementarity relation is logically prior and serves to *explain* the limitations expressed by the uncertainty relations.

At this point I would emphasize a well-worn theme. Bohr's interpretation of quantum mechanics is not a subjective one. By limitations on knowledge in the above discussion Bohr refers to results of experiments, not our own *experience* or *conscious awareness of these results*. As Bohr puts it in his (1963) (p. 3):

> All unambiguous information concerning atomic objects is derived from the permanent marks . . . left on the bodies which define the experimental conditions . . . The description of atomic phenomena has . . . a perfectly objective character, in

the sense that no explicit reference is made to any individual observer.

IV. The Role of Classical Concepts

We now reach a theme constantly emphasized by Bohr, the *necessity* for describing experiments and their results using classical concepts. This is often given a Kantian interpretation. See, for example, Honner (1982). But Folse in his (1985) argues persuasively that Bohr is concerned with the conditions for unambiguous communication of our experience, not with the conceptual framework constitutive of that experience. To return to Bohr, in a familiar quotation (Bohr (1963) p. 3):

> The decisive point is to recognize that the description of the experimental arrangement and the recording of observations must be given in plain language, suitably refined by the usual physical terminology.

This sort of quotation is often given the following gloss. The experimental arrangement of diaphragms, shutters, slits, photographic plates and so on, *must* be described in the language of classical physics insofar as it is used to do the observing. If, however, such apparatus is itself being observed, then it must itself be subject to quantum-mechanical description, i.e. only a limited applicability of classical concepts such as position and momentum is permitted as specified in the uncertainty relations.

In other words the type of physical description we use depends on the *role* of a macroscopic object such as a diaphragm. We shall refer to this thesis as *epistemological relativism*. We quote Bohr again ((1963) pp. 3–4):

> The essentially new feature in the analysis of quantum phenomena is, however, the introduction of a *fundamental distinction between the measuring apparatus and the objects under investigation*. This is a direct consequence of the necessity of accounting for the functions of the measuring instruments in purely classical terms.

Indeed Bohr emphasizes the latitude in where we place the subject/object division—often referred to as the *moving cut* between what is observed and what is doing the observing. Thus in his (1934) p. 54: "It is a question of convenience at which point the concept of observation involving the quantum postulate with its inherent 'irrationality' is brought in." I will now make some comments on this familiar gloss.

The moving cut cannot be placed totally arbitrarily. For example, in the Stern-Gerlach measurement of the spin component of an atom, if we regard the first stage of the measurement chain as correlating the spin component with the transverse momentum of the atom, it is clearly not Bohr's intention to say that, because the atomic momentum is measuring the atomic spin, the former *must* be described classically.

Bohr recognizes that *must* has to be reconciled with *can*. So in Bohr (1963) (p. 91) we read:

> This demand [i.e. describing the experimental arrangement by classical physics] is fulfilled by using as measuring instruments bodies like diaphragms, lenses and photographic plates so large and heavy that, notwithstanding the decisive role of the quantum of action for the stability and properties of such bodies, all quantum effects can be disregarded in the account of their position and motion.

So the moving cut must be located somewhere within a sequence of bodies where classical description is *approximately* possible. Notice the reference to position and motion—there is no mention here of momentum. This is an important point that seems usually to be overlooked. When Bohr says the experimental arrangement is to be described using the concepts of classical physics, it is often assumed that the description is *entirely* classical, i.e. that *all* the classical mechanical concepts apply.

But taking the case of space-time coordination, for example, consider this quotation (Bohr (1958) p. 44):

> The position and motion of the diaphragm and the shutter have . . . been assumed to be accurately coordinated with the

> space-time reference frame. This assumption implies, in the description of the state of these bodies, an essential latitude as to their *momentum* and *energy*, which need not, of course, noticeably affect the *velocities*, if the diaphragm and shutter are sufficiently heavy. (My italics.)

The point being made seems to be this. So far as position and velocity are concerned sufficiently heavy bodies used to identify space-time coordination obey classical physics, but this is not true in respect of their momentum.

In terms of the uncertainty relations

$\Delta pr\ \hbar/\Delta x$ and $\Delta v \sim \hbar/\Delta x$

For a given Δx, however small, Δv can be made as small as we please by increasing m sufficiently, but this does not help in reducing Δp.

Two points should be noticed:

(1) To correlate with an objective fixed space-time frame it is no good just having $\Delta x = 0$ at a given instant of time. We need $\Delta v = 0$ (effectively) so that the spatial coordinates *remain* definite as time develops.
(2) If we actually let $m \to \infty$ and *then* let $\Delta x \to 0$ we would get $\Delta v \to 0$ and $\Delta p \to \infty$. But notice that, if $m \to \infty$, Δp becomes a totally metaphysical (i.e. unmeasurable) quantity, because one can *only* measure Δp by measuring Δv first. But since $\Delta v = 0$, $\Delta p = m\,\Delta v = \infty \times 0$ is left quite indeterminate.

The upshot of these quotations seem to be that the apparatus is not described totally by classical physics, but only by that part of it relevant to its *fulfilling the function for which it is being used*—space-time coordination in the examples under discussion.

Bohr makes this point very clear in the following quotation. In his (1937) p. 293 he says: "... *Instruments according to their purpose* shall allow the unambiguous use of the concepts necessary for

the description of experience." (My italics)
Or again in his (1963) p. 5:

> [In] an experimental arrangement involving the transfer of momentum and energy to bodies such as fixed scales and synchronized clock . . . [this transfer] cannot be included in the description of their functioning if *these bodies are to fulfill the rôle of defining the reference frame.*" (My italics.)

So at this point we are faced with two possible interpretations of Bohr:

(1) The measuring apparatus is described by *all* the classical concepts, i.e. is subject to classical physics.
(2) The measuring apparatus shall allow those concepts in their description necessary for them to fulfill the function for which they are being used, position and velocity for spatial location, momentum for momentum determination and so on.

We will now proceed to criticize these two views in turn.

V. The Epistemological Relativity of the Classical Description

Firstly we look at the alternative (1). This is adopted by Scheibe, for example. He relies on the following sort of quotation: "The description of the properties and manipulation of the measuring instruments [based] on purely classical ideas *implies the neglect of all quantum effects in that description* . . . " (Bohr (1939) p. 19, my italics). That certainly sounds like position (1).

But Bohr goes on to say:
" . . . and in particular the renunciation of a control of the reaction of the object on the instruments more accurate than is compatible with the [uncertainty] relation." So Scheibe suggests that the complete classical description is a *misdescription* and goes on to say (Scheibe (1973) p. 27): "The error thus committed in respect of the apparatus is then held responsible for the impossibility of precise surveillance of the interaction."

But if the complete classical description is a misdescription, one is tempted to turn to position (2) and say that the correct description of the apparatus is also quantal. Of course Bohr would agree with this if the apparatus is made the object of investigation in the interpretation of Bohr which we have characterized as the epistemological relativity of the classical description.

However, Bohr's usual arguments in this connection are concerned with trying to survey the momentum of a diaphragm with a fixed position. But suppose we observe the position of a diaphragm which already has a fixed position.

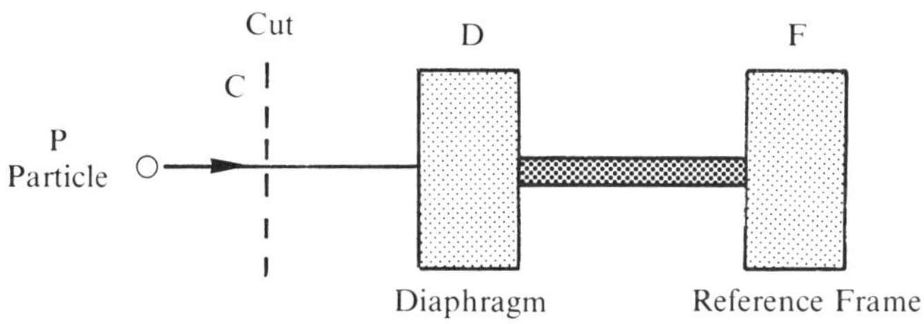

First the diaphragm is being used to locate the position of the particle *P* by being bolted rigidly to the reference frame *F*. (Imagine the diaphragm to be coated with a photographic emulsion or to have a narrow slit cut in it so as to fix the transverse position of *P*.) Now "measure" the position of *D* by attaching it to a block *F′* which is rigidly attached to *F*.

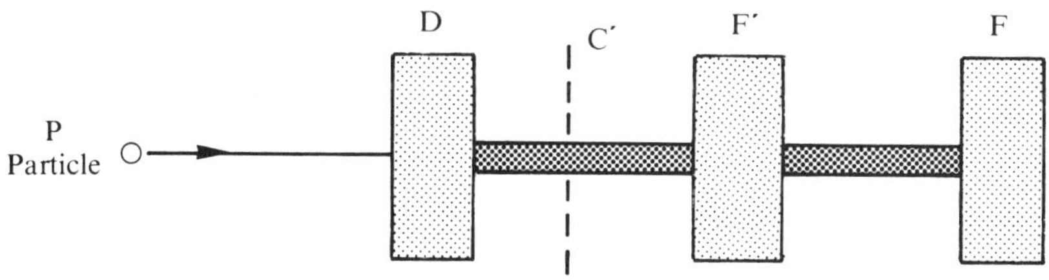

Should the description of *D* change when we insert *F′*?, i.e. move the "cut" from *C* to *C′*. In particular does *D* suddenly acquire a sharp momentum, when we take out *F′*? This sort of example is never discussed by Bohr.

But it does at the very least raise serious doubts about the credibility of view (1).

If something is never observed, our choice of description becomes arbitrary and metaphysical. But, for Bohr, there is no ultimate apparatus *F*. *Everything* is relativized. There is an infinite regress, that does not terminate for Bohr in human consciousness (pace von Neumann, Wigner, etc.) and which precludes *in principle* cosmological applications of quantum mechanics (this after all was the motivation of Everett and Weeler in their so-called many-worlds interpretation.)

Let us now look at the second alternative.

VI The Quantum Mechanics of Macroscopic Objects

The epistemological relativity thesis is a deep and interesting one, but extremely difficult to make sense of. I do not say it is a hopeless task, only that I do not myself know how to do it.

The second view is at first sight more promising, that, quite independently of where we locate the "moving cut," all objects, macroscopic or not, are described quantum-mechanically not classically, but that the partial descriptions should always employ classical concepts. Bohr famously says ((1934) p. 54): "An independent reality in the ordinary physical sense can neither be ascribed to the phenomena *nor to the agencies of observation.*" (My italics.)

As we have remarked Bohr appears to believe that quantum mechanics tends to a classical limit as $m \rightarrow \infty$. But we must not actually proceed to the limit, because in so many of his thought experiments one can measure Δp for diaphragms, etc. The fact that quantum mechanics *does* apply to macroscopic phenomena is nowadays commonplace in the theory of superconductivity, etc.

In terms of the quantum-mechanical formalism what Bohr seems to have in mind is that if a heavy body starts in an approximate eigenstate of position it will *remain* in such a state indefinitely. The usual analysis of dispersion of wave-packets is relevant here. As $m \rightarrow \infty$ the dispersion is infinitely slow.

But what Bohr does not seem able to appreciate, on the view we are now discussing, is that if a macroscopic object gets into a superposition there is no way of getting out of it. Discussion of spreading of wave-packets is quite irrelevant to that problem.

It is of course the famous problem of measurement in the formal quantum-mechanical treatment, highlighted in the Schrödinger Cat Paradox.

VII. Conclusion

The two views we have discussed correspond in the formalism to the two-motion theory of measurement with the projection postulate, and the one-motion treatment, only contemplating Schrödinger time-evolution.

Again the difficulties we have considered reappear in the formal account. In the two-motion theory it becomes entirely arbitrary and mysterious as to where and how the second motion takes over. In the one-motion theory measurement never happens.

Nothing that Bohr says about complementarity seems to help with resolving this basic dilemma.

But let us suppose, for the sake of argument, that the epistemological relativity thesis could be properly worked out. Then we must address the question, the title of this paper, what should the rôle of complementarity be for the future of physics?

Here we must be fair to Bohr. He never claimed that complementarity was the last word. In his (1937) (p. 294) Bohr says: "We must be prepared for a more comprehensive generalization of the complementarity mode of description which will demand a still more radical renunciation of the usual claims of so-called visualization."

But nevertheless it is clear that, although allowing for generalizations, Bohr thought that in complementarity he had made a profound philosophical discovery in the context of atomic physics but which could be applied in a much wider context of biology, psychology and sociology.

But that is where the danger lies. Metaphysics certainly guides the construction of theories in physics, but it must never be allowed to become a straight-jacket. That way lies stagnation, the myth of final achieved metaphysical truth.

We must always be prepared in physics to look for new ways of doing physics, and that inevitably means new ways of doing metaphysics.

NOTES

1. The paper referred to is Bohr and Rosenfeld (1933).
2. For a recent survey of this work see Bohm and Hiley (1985).

REFERENCES

Bohm, D. and Hiley, B. J. (1985): "An Ontological Basis for the Quantum Theory," paper presented at the International Conference on Microphysical Reality and Quantum Formalism, Urbino, Italy.

Bohr, N. (1934): *Atomic Theory and the Description of Nature* (Cambridge: Cambridge University Press).

Bohr, N. (1937): "Causality and Complementarity," *Philosophy of Science, vol. 4*, pp. 289-298.

Bohr, N. (1939): "The Causality Problem in Atomic Physics," in *New Theories in Physics*, (Paris: International Institute of Intellectual Cooperation), pp. 11-45.

Bohr, N. (1958): *Atomic Physics and Human Knowledge* (New York: Wiley).

Bohr, N. (1963): *Essays 1958-1962 on Atomic Physics and Human Knowledge* (New York: Interscience).

Bohr, N. and Rosenfeld, L. (1933): "Zur Frage der Messbarkeit der elektromagnetischen Feldgrössen," *Mat.-Fys. Medd. Dan. Vid. Selsk., vol. 12* No. 8.

Folse, H. J. (1985): *The Philosophy of Niels Bohr: The Framework of Complementarity*, (Amsterdam: North-Holland).

Honner, J. (1982): "The Transcendental Philosophy of Niels Bohr," *Studies in History and Philosophy of Science, vol. 13*, pp. 1-29.

Petersen, A. (1963): "The Philosophy of Niels Bohr," *Bulletin of the Atomic Scientists*, vol. 19, pp. 8-14.

Redhead, M. L. G. (1983): "Quantum Field Theory for Philosophers," in P.D. Asquith and T. Nickles (eds.) *Proceedings of the 1982 Biennial Meeting of the Philosophy of Science Association*, vol. 2, (East Lansing: Philosophy of Science Association), pp. 57-99.

Rosenfeld, L. (1953): "Strife about Complementarity," *Science Progress, vol. 41*, pp. 393-410.

Scheibe, E. (1973): *The Logical Analysis of Quantum Mechanics* (Oxford: Pergamon).

SYMMETRIES OF PERSONAL PROBABILITY KINEMATICS

(with an Appendix by John Collins)

BAS C. VAN FRAASSEN

EPISTEMOLOGY is the theory of knowledge and opinion. For empiricism, fascinated with the erosion of certainty in our view of nature, opinion is the more important topic.

Personal probability is one model of opinion. Or rather, to be more accurate, it is the main ingredient in several competing, but related models. I wish to present and discuss some developments in this approach, especially ones pertaining to the treatment of change of opinion. As always in model construction, attention to symmetries provides an important conceptual tool, and I would like to convey the role it plays here.

To begin, I shall locate this approach to some extent in general epistemology. Then I shall concentrate on very specific questions about how we can rationally change our opinions, in response to experience. I shall end by outlining some open problems that I commend to your attention.

I. Two Epistemological Traditions

In most discussions of opinion, only three epistemic attitudes are discussed: belief, disbelief, and neutrality. Richard Jeffrey has explained how there has always existed a minority tradition, an alternative epistemology, in which opinion is represented as having many gradations and forms. Thus he traces back our current recourse to probabilities in epistemology, not merely to Savage and Carnap, or to Ramsey and de Finetti before them, or to Jevons and De Morgan in the nineteenth century, Bayes in the eighteenth, Huygens and Pascal in the seventeenth, but even to Carneades in late Greek antiquity. That this has always been a minority view must be admitted. It still is, in part perhaps because many philosophers are tempted to honor the humanities by ignoring the sciences (as well as vice versa). I shall therefore do my best to explain this approach in a way that locates it with respect to the most general epistemological concerns.[1]

II. The Differences

Just how different is this alternative epistemology? I think it is less different in its main concerns than appears at first—but more different than it seems even to some of its adherents, in the ways of securing those concerns. The most obvious difference is the one I mentioned to begin: the description of the varieties of opinion. The main tradition concentrates on three: full belief, full disbelief, and the neutral attitude of "having no opinion." (This third attitude is very important, and naive probability models of opinion tend to exclude it, which they should not. To have no opinion about whether the accused is guilty is *not* the same as saying that he seems just as likely to be guilty as innocent. That is a definite sort of opinion, and not lack of opinion.) Jeffrey calls a state of opinion which can be fully characterized by means of these three categories, *dogmatic*. The alternative tradition concentrates on a much richer set of attitudes, which can be expressed in, for example, the following forms:

(A) It seems likely (certain, unlikely, very likely) to me . . .
(B) It seems as likely as not . . . It seems twice as likely as not . . .
(C) It seems likely that . . . on the supposition that . . .
(D) It seems more likely that . . . than that . . .
It seems twice as likely that . . . as that . . .

In (A) we have a plethora of qualitative degrees of some sort, enlarged to an infinity of numerically precise ones in (B). Conditional and comparative variants of these appear in (C) and (D).

The second difference is that the relationship between opinion and the world is more complex. If I express the belief that it will rain, it then rains or does not, and my opinion was correct (true) or incorrect (false) accordingly. If I say that rain seems very likely to me, it then rains or doesn't; and what does that mean about how "good" my opinion was?

The third difference lies in the epistemologically relevant relations *between* the opinions. In the main tradition, this had two aspects: the logical evaluation in terms of consistency and inconsistency, and quite independently, the evaluation of *reasons for belief*. In the new, both differ. The logical evaluation must be extended to such forms as (A)—(D) above. (Reflecting on the preceding paragraph, we don't expect the familiar semantics of reference, truth, and extension to do the job.) Secondly, the whole idea of reasons for opinion—let alone induction or "inductive support"—becomes unintelligible, as far as I can see, at least in the stricter forms of Bayesianism, and perhaps everywhere. My reasons for thinking that (Yes!) have really to do with the fourth difference, which concerns rational change of opinion, and which is the main topic of this paper.

III. How Does Probability Come In?

You already think and speak probabilistically if you are able to express your opinion in forms (A)—(D) above. They each have a jargon equivalent, like this:

(A) My subjective probability is high (= 1, low, very high)
(B) My subjective probability = ½ (*or*: odds are 1:1)
My subjective probability = ⅔ (*or*: odds are 2:1)

and so forth. A *probability function* assigns 1 to tautologies, 0 to contradictions and real numbers in between to all others in a coherent fashion. You already work intuitively with them without the jargon. To gain equal facility with the numerical calculation, all you need is to have the right model in mind. This model is the muddy Venn diagram.

You represent the propositions by areas on a Venn diagram, and their probability by mud heaped on them. Call the total mud present 1 unit; the proportion of mud on the proposition area equals its numerical probability. This model is better than any axioms or rules for calculation, and it does equally well. (There are in fact deep theorems to show that we have here the most general model of probability theory, provided the mud is so fine as to be continuous.)

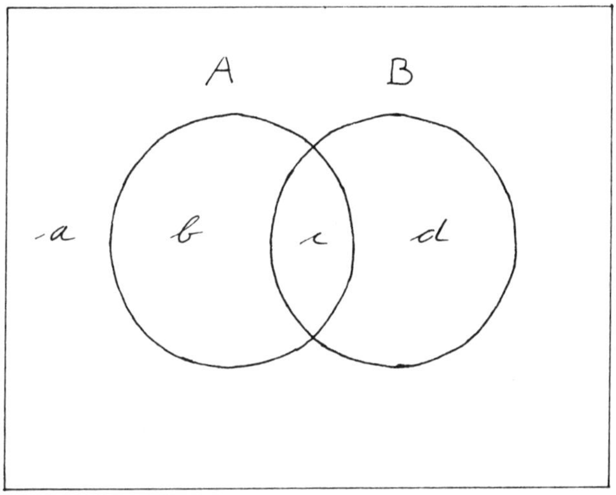

It may be objected that pictorial model or no, our computational abilities with numbers leave something to be desired. Policing our own state of opinion for logical coherence is therefore in practice not easy, and is even in theory in some respects entirely beyond human capacity. However, the same applies to logical computation, and the problem posed is therefore not a difference between the main and alternative epistemological traditions.

It may also be objected that we do not have numerically precise comparative opinions for all the pairs of propositions we can contemplate. This is correct, indeed, I can't always express such an opinion as that rain seems, or seems not, as likely as snow. *So our subjective probabilities are vague.* This is the answer that was given by Richard Jeffrey among others. The sophistication it introduces requires us to reflect on vagueness, which has some associated technical difficulties. I will leave these aside except to say that vague opinion is to be understood in such a way that we can continue to use the muddy Venn diagram to check for coherence. It would not be coherent for instance to profess vague subjective probabilities of say 0.5 ± 0.1 for (rain or snow) and 0.8 ± 0.1 for (rain and snow).

To reason vaguely, requires an understanding of how one reasons precisely. Consider the propositions:

> (*) John is very tall; Peter is taller than John; Peter is not very tall.

What is wrong with this? Well, "very tall" is vague. But if we ask why, precisely, the first statement is true, we will be told that John's height is greater than some number x. The second statement then entails that Peter's height is no less than x, and this refutes the third statement just as much as the initial datum established the first. So the vague reasoning that regards (*) as inconsistent, rests on a grasp of the validity of the precise argument form: $t > x$; $t' > t$; therefore $t' > x$. Hence we can, and should, simultaneously insist that our opinions are vague and gappy and not numerically precise—but that to understand their structure we should study precise, full, and numerically defined opinion, however idealized that may be.

IV. Harman's Objections

Gil Harman has objected to the subjective probability model of opinion as follows. The memory capacity of the brain might well be outstripped by the assignment of numerical probabilities to all sentences in even a quite simple language. It should be pointed out here that these numbers cannot be assigned by recursion, starting with an assignment to the atomic sentences. The reason is that how likely a conjunction is, for example, is not determined entirely by how likely its conjuncts are. For vague probabilities it is even worse he says, since to remember them you must typically store *two* numbers for each sentence (e.g., 0.8 and 0.2 for 0.8 ± 0.2).

This objection certainly had me stumped for a while, and although Jeffrey did have several ideas that could ameliorate the difficulty, I did not feel satisfied. But I think now that I can describe a model of opinion which is not prey to Harman's objection, and yet preserves the essential role of subjective probability. It is this. My state of opinion is characterized first and foremost by a set of judgements—of such forms as (A)—(D); these are the judgements which are *mine*. Next we define a relationship of *satisfaction* between probability functions and judgements. For example:

(4.1) p satisfies the judgement that A seems less likely than B, just in case $p(\text{A}) < p(\text{B})$.

The set of all probability functions which satisfy *my* judgements, I'll call my *representor*. My state of opinion is *coherent* exactly if this set is not empty. Now there are some judgements which I could not add without becoming incoherent, and some that I could add without changing my representor at all. We have here the obvious generalizations of logical evaluation of belief. If a judgement, for example, is satisfied by all members of my representor, then I am committed to it, as opposed to its contraries, on pain of incoherence. So I'll say it is *entailed* by my judgements. Here is an example:

It seems twice as likely as not that A . . .
It seems at least likely that B as that not A . . .
Therefore: my subjective probability for B is no lower than ⅓ (or, equals ⅔ ± ⅓).

In this way I am committed to many numerical (though vague) probability judgements. The logic of vague judgement does not have any great mystery about it. Finally, my opinion can be stored in memory as a finite set of judgements. However, as Harman has pointed out, we need to investigate whether the result of conditionalizing such a vague state of opinion on a given proposition will always yield a new one also characterizable by a new finite set of judgments.

V. The Deliverances of Experience

Let us now turn to change of opinion. There I want to concentrate solely on change in response to experience, and its rationality.

(In other words, I leave out processes of deliberation, theoretical innovation, conjecture, conceptual change, or whatever else there be. That there is room for this in the personal probability model, I have argued in my "Belief and the Will.") *There is a general obstacle.* Suppose I were to write a recipe, or book of recipes, that would tell you how to amend your opinion rationally in view of your experience. The recipe would begin with a description of a state of opinion, and then a description of experience, or of the deliverances of experience, and then prescribe new opinions. Now you could use this recipe to evaluate how well someone else was doing, if you thought you knew what his experience and opinion were. But if you tried to use it on yourself, you would first have to describe your experience. And that is already a response to experience, and would be expressed in terms of a whole new set of judgements that are yours. So you would already have done a great deal of the job that the recipe is meant to guide. You cannot really step out of yourself and compare the representation with what it represents.

I didn't describe this problem either because I think there could be a serious use for recipes here, or because I want to discuss foundationalism of any sort. I wanted to bring out instead the need for a model of experience, if we want to continue the epistemological story. And I do not mean a physiological or psychological model, because the focus of interest is not so much on how things happen, as on the rationality of the response that it must somehow be possible to evaluate. The model must pertain to phenomena presumably reported in such utterances as : "I saw a flying saucer last year; and ever since I have been a firm believer in reincarnation."

There are three models. The first was inherited I think from the main tradition. It says that in experience, some proposition E is received as evidence (or, it is taken as evidence), the subject becomes immediately totally certain that E, and adjusts his opinions accordingly. The rule for adjustment is *Simple Conditionalization*, to which I will return below. Perhaps also this model is perfectly suitable to another source for the Bayesian tradition: the working statistician, after all, is paid to accept certain propositions as data, not to question them but to use them as input for his calculations. But as a model for experience it is a bit simple-minded. I call it the *Revelation Model* because in it, experience speaks with the voice of an angel and gives you new total certainties.

A second model was described by Hartry Field in his article about Jeffrey. He takes it as an article of faith that any epistemology must be compatible with materialism. So the input is a physical stimulus; the opinion is in physical storage, and is physically modified. I call this the *Robot Model*. Here the input is not a proposition, obviously, and need have nothing to do with propositions directly. Nor is there any question of an evaluation of the rationality of the response. It may be possible to speculate about general features of the mechanism, perhaps with an eye to survival value.

The third model, which I wish to propose, also does not take the deliverances of experiences to be propositions, though more intimately connected with them. There is undoubtedly some level of response which we cannot criticize effectively in ourselves, even if we think that it too might be erroneous in some sense. Let me call this the primary response. But I do not think that this primary response must already be a judgement, let alone a new total certainty. It is instead, I think, the acceptance of a constraint on what your opinion (henceforth) should be. Thus the deliverances of experience are not propositions, but commands. A limiting case is possible: the command to become totally certain that E. If you accept that as a constraint, what happens next must be what happens next in the revelation model. But many different sorts of constraints are possible—for example, to raise a subjective probability a little, or accept new odds, or a new conditional probability, or indeed anything that could constrain opinion. Because of the crucial roles of the terms "accept," "constraint," "command," it seems natural to call this model a *Voluntarist* one.

Here questions of evaluation can and do arise. Suppose you accept a constraint on your actions by saying you will post a letter for me. I can later evaluate (a) whether your actions satisfied the accepted constraint, and (b) if so, how well you did. If for example you dropped the letter in the mud and got it to the mail box several days later, you satisfied the constraint, but not optimally so, in some interesting respects. Similarly for adjustments to opinion, made as secondary response, when certain constraints have been accepted.

In the remainder of the paper I will discuss examples of primary response, and putative rules to govern the corresponding secondary response.

VI. New Certainties: The Voice of an Angel

Suppose I go for a walk in the garden and come away absolutely convinced that a flying saucer has landed there. I reconstruct this as follows: I had originally a certain state of opinion, but accepted the constraint to become certain of this new proposition E, and adjusted my opinion accordingly. There is, as I mentioned, a rule for this adjustment, *Simple Conditionalization*. It is easily explained in terms of the muddy Venn diagram.

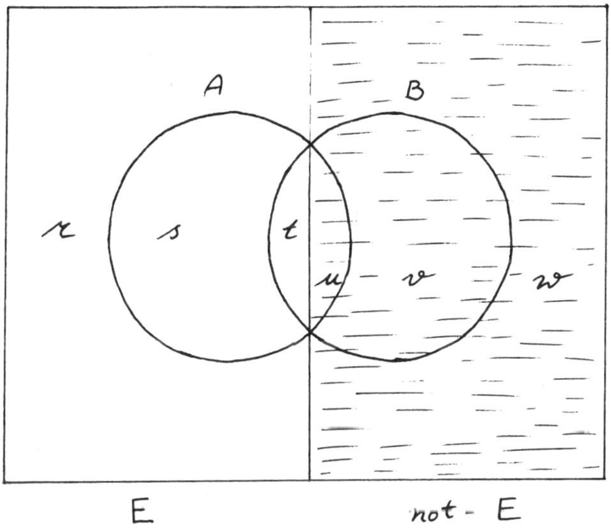

You simply wipe away all the mud on the area representing not-E. This has two effects: it raises the probability of E to 1 (for all the mud remaining is on E) and keeps the odds between propositions that entail E the same (for the mud "inside" E was not disturbed. This is a complete description of the rule. If probability function P is changed in this way we say it is *conditionalized* on E and call the result P_E or $P(- \mid E)$.

It is also possible to state this rule in algebraic form (which follows logically from its description in terms of the mud model): $P(X \mid E) = P(X \text{ and } E)/P(E)$. In either form it is obvious that the rule cannot apply if the prior probability $P(E)$ equals zero. In that case wiping away the mud on (not-E) would remove all mud, thus destroying the model.

So much for the geometric and algebraic descriptions of the rule—*but is it right*? This question of justification is a very fair one. Ian Hacking, writing in 1967, noted that Bayesians took this rule for granted and he called it the *Bayesian Dynamic Assumption*. Textbook presentations tend to darken counsel, as usual, by suggesting that the meaning of "conditional probability" is "the probability you would have if you had learned that." This cannot help because even if it were the meaning, that meaning has logically nothing to do with my present odds for (X and E) to (not X and E)—which is the information conveyed by the number $P(X \text{ and } E)/P(E)$. In 1976 Paul Teller reported an argument due to David Lewis, attempting to justify the rule by a so-called Dutch Book argument. That has an interesting history (see Shafer, my paper on Shafer, and "Belief and the Will"). But Teller also presented an argument of his own based on considerations of continuity and relevance, later improved in his joint paper with Arthur Fine. I think now that the rule has an all-but-completely a priori justification, by means of a symmetry argument (see further the end of the next section).

VII. New Degrees: Roses by Candlelight and Lying Spies

In 1965, Richard Jeffrey created the new subject of *probability kinematics*. He did this by describing different sorts of changes in opinion, asking for a theoretical description, and proposing a rule that generalizes simple conditionalization. This new rule has since been generally known as *Jeffrey Conditionalization*.

Here is an example: I walk through a room, in which I glimpse roses on a table lit by candlelight. I had expected the scene, had a prior opinion that the roses would, as likely as not, be red. Now I am more inclined to think they are red . . . it now seems twice as likely to me as not that they are red. How should I adjust my total state of opinion "accordingly"?

A bit later or at roughly the same time as Jeffrey was thinking about this, I have been told, Wade Edwards and his collaborators were discussing a *prima facie* different problem. The rule they came up with is formally the same as Jeffrey's. Their problem was this: we have a spy in Iran, and he sends us regular reports, but he is non-reliable, he tends to lie a little. Reading his reports, what constraints shall we place on our posterior opinion? There is an obvious analogy with Jeffrey's case, a suggestion that we think of our senses as Plato did, as lying spies in the garden of earthly delights. Perhaps my eyes were saying "red" when I walked through that room, but they did not speak with the voice of an angel, and my response was only to become more inclined toward, *not* certain of, the proposition they spoke.

Let us restate the problem in an even more schizophrenic mode. There are two spies, Cain and Abel; they are twin brothers, and their reports are always each other's negation. But we consider Cain twice as likely to speak the truth as Abel. Now we can say this: if I accept Cain's reports totally, I shall conditionalize on them; if I accept Abel's totally, conditionalize on the negations of Cain's.

Neither of these brothers speaking like an angel, I shall make my own opinion a proportional mixture of theirs.

This idea of a *mixture* of two states of opinion works perfectly in the alternative epistemology. In the dogmatist oversimplification, the best one could do would be to take what is implied by both Cain's and Abel's reports—i.e. tautologies only.

This is the complete description of the rule of Jeffrey Conditionalization.

I can now also address simultaneously the justification of Strict and Jeffrey Conditionalization. Let us first state carefully when the rules apply. We are given a *prior* state of opinion, represented by probability function P. In addition we have a single constraint which must be satisfied by the *posterior* state of opinion. If the latter is to be represented by a probability function P', the constraint takes the form $[P'(E) = y]!$ for $y < 1$. The task is to find a rule which transforms P into P'. We are allowed to make no special assumptions. But the rule too may not utilize any further information; otherwise we have solved the wrong problem. This reflection we can translate into what physicists call a *symmetry requirement*: it spells out what is to be meant here by "essentially the same" in the adage, "If two problems are essentially the same they must receive essentially the same solution as well." On this basis, we can now in fact prove the uniqueness of the rules as given:

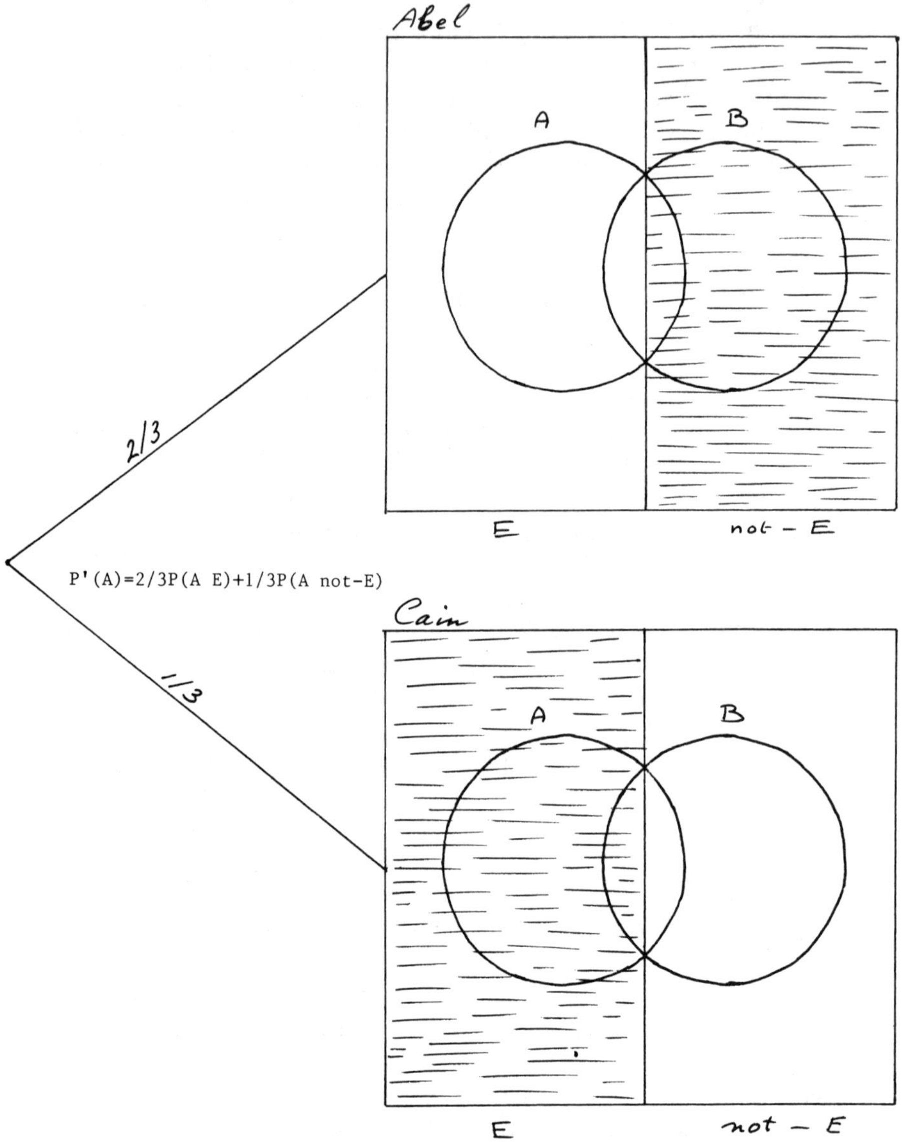

The rule of Jeffrey Conditionalization (and its special case Simple Conditionalization) does respect the symmetries of the problem, and no other rule does (see Hughes and van Fraassen 1985; van Fraassen, 1986).

That does not mean that a person must, to be rational, live by this rule. To prove this result it was necessary to say when two such problems are essentially the same, and this amounts to an assertion about which the sole relevant factors are. (In this case: the constraint, and the prior probabilities alone.) I have no immediate argument to rule out as irrational someone who asserts that rational procedure must let the exact change of opinion depend also on the weather, mood, breakfast, or phase of the moon. Nor can I rule out as irrational the decision not to follow *any* rule in certain cases, but to make a leap of faith. Indeed, I am inclined to the traditional view that to accept hypotheses going well beyond one's evidence is in itself within rational procedure (see my 1984, 1985). However, this is certain: Strict and Jeffrey Conditionalization are the uniquely favored *rules* for their proper domain of application.

VIII. Levi's Objection: A Simulated Horse Race

Also writing in 1967, in a review of Jeffrey's book, Isaac Levi raised certain objections to the idea of probability kinematics. One of these has variants for all continuations of the program as well. Surely one's opinion, formed in response to a succession of experiences, should not depend on the *order* in which these experiences occur? That is obvious if the experience fits the Revelation Model. Simple Conditionalization successively on E_1 and E_2 is the same as doing it on (E_1 and E_2)—so there order does not matter. But Jeffrey Conditionalizing, with successive constraints $[E_1 \Rightarrow 0.7]!$, $[E_2 \Rightarrow 0.8]!$ for example, clearly does depend on the order. (Think of the limiting case $1 = 2$.) And so it should; but if we have here a correct description of adjustment of opinion in response to relevant experience, we have a bit of a paradox. Two persons, who have the same relevant experiences on the same day, but in a different order, will not agree in the evening even if they had exactly the same opinions in the morning. Does this not make nonsense of the idea of learning from experience?

I would like to give two responses. First, let us ask whether experience can contradict itself? Surely not: what my experience is at two different times, is a priori unrestricted—what happens to me is up to nature, so to say, and there are no a priori bounds on nature. (This is an empiricist speaking.) But what about responses to successive experiences? Do we have any reason to suspect irrationality if those are incompatible? I think not; else why ever look twice? It is along these lines that I understand Jeffrey's own response. That is that a person who wishes to raise the probability of E_1 to 0.7 and a bit later that of E_2 to 0.8, has the choice of imposing the joint constraint $[E_1 \Rightarrow 0.7$ and $E_2 \Rightarrow 0.8]$! at the later time, provided the two constraints are logically compatible. And if they are not compatible, he clearly *wants* to discard the previous judgement.

This response has two problems. First, Jeffrey's rule does not tell you how to impose the joint constraint. This requires a rule going beyond his; and such rules raise questions of their own, which I shall discuss below. The second problem is that the incompatibility is not a matter of simply discarding past judgements. Suppose E_1 and E_2 are mutually incompatible. Then they cannot have .7 and .8 at the same time; and which you impose later makes a difference to what everything gets. As an example, suppose I have the situation of Display No. 1.

Display No. 1

		E_1	E_2	not (E_1 or E_2)	
P:		.1	.8	.1	impose $[E_1 \Rightarrow .7]$!
P':		.7	$\frac{8}{9}(.3)$	$\frac{1}{9}(.3)$	
	i.e.:	$\frac{21}{30}$	$\frac{8}{30}$	$\frac{1}{30}$	impose $[E_2 \Rightarrow .8]$!
P'':		$\frac{21}{22}\left(\frac{2}{10}\right)$	.8	$\frac{1}{22}\left(\frac{2}{10}\right)$	

Here you see clearly that going back to the original value for E_2, for instance, does not get you back to the original state of opinion. If the constraints had been imposed in opposite order, the first would have effected no change at all, and P' would have been the end result.

So we need a more sophisticated response. I think we should momentarily forget the technicalities, and ask: What does happen to learning from experience? Has this become impossible? Suppose a person *learns*, in the sense that he successively accepts constraints, each of which correctly represent the real, objective probabilities. Will his opinion become more correct, or not?

This question could be investigated experimentally, to some extent. I decided to simulate an experiment on a computer. A man wishes to bet on a horse race; on each day preceding the race he is allowed to study the horses. Suppose that on each day D_n he updates his probability that a certain horse H_n will win to P_n, using Jeffrey's rule. Suppose also that the correct or objective probabilities are indeed $P_1, \ldots, P_n, \ldots$ Should we expect his opinions on the day of the race to be closer to the correct probabilities than they were initially? For simplicity I used three horses, and asked the question in this form: how many days does it take for his subjective probabilities to come within a preset amount of the objective ones? Display No. 2 indicates some results (stated in terms of odds rather than probabilities).

Display No. 2

			Degree of Approximation			
	Prior	Target	1/100	$1/10^4$	$1/10^8$	$1/10^{10}$
(A)	1,1,1	$1,10^3,10^6$	2	2	2	3
(B)	1,1,1	1,10,100	2	4	6	8
(C)	1,1,1	1,1,20	2	5	8	11
(D)	3,3,4	1,2,7	3	5	11	13
(E)	7,2,1	1,2,7	5	7	11	14
(F)	5,6,7	6,7,8	1	5	11	15
(G)	5,6,7	7,8,9	1	5	12	15
(H)	1,5,25	1,2,4	3	6	11	17

There is no doubt that this person is learning, and quickly. He gets to within one percent of the right probabilities in a few days, and to within an astronomical degree of approximation within about two weeks. Not only that, there is a definite convergence, and it is not true that successive experiences "undo" the preceding learning process to a radical extent. Of course I assumed that the *input*—the imposed constraints—were exactly correct. But while such acute perceptions may be lacking in reality, any lack of overall success will not be due to the rule being followed.

The experiment does raise a *question*, to which I have no answer. The speed of approach is clearly different in examples (A)—(H). That speed must be determined by some relation between prior and target. But I don't know of any quantity that can be used as an indicator of this speed of approach. I conjecture that the convergence we see in the examples is not accidental, but can be proved relatively easily; the proof may yield some suggestion for what such indicators could be.

IX. The General Problem Reformulated: Symmetry Again

In 1965 Jeffrey had formulated a simple, new problem: in the roses-by-candlelight case, where Simply Conditionalization is not applicable, what should one do? And as we saw, his proffered solution has a very special status, for it can be justified a priori. As we also noted, there may be cases in which this result still gives us no guidance. This can be for one of two reasons:

(a) We may want our posterior opinion to depend on other factors besides the prior opinion and the given, simple constraint; or
(b) We may wish to impose another constraint, of more complex form.

Reason (a) is always with us, and rightly keeps us from conceiving epistemology as in principle a special sort of arithmetic or other mechanical procedure. But we should also consider the case—presumably "normal" in the sense that deliberation and theoretical innovation and creativity must be the exception rather than the norm in our daily life—in which reason (a) is absent.

This leads us to the problem of *general probability kinematics*: find rules that transform priors, subject to given constraints, and for which the prior and constraint are the sole relevant factors.

In a paper written jointly with R.I.G. Hughes, I have presented the symmetry argument outlined in section VII above, in another key. It turns out to be sufficiently strong to demonstrate the following general result. Suppose that we are given constraints on the posterior probabilities $P'(A_1), \ldots P'(A_n)$ defined with reference to a partition $(A_1, \ldots, A_n)$ of the space of possibilities. Then any rule that transforms a prior P into such a posterior P' must in effect proceed in two steps:

(i) Assign exact new probabilities to $A_1, \ldots, A_n$
(ii) Treat the outcome of step (i) as input for a Jeffrey Conditionalization

This result is very helpful, because it means that if we are given a constraint of a new and strange sort, like:

> Change your opinion so that the probabilities for rain and snow become equal.

we know that we need not worry about fine subdivisions like (rain and cereal for breakfast), (rain and scrambled eggs for breakfast), etc. For step (ii) must always be carried out in the same way. With this in hand, let us begin to investigate constraints of more sophisticated form.

X. New Conditional Probabilities: The Judy Benjamin Problem

I have already mentioned one task which cannot be done by Jeffrey's rule (alone): imposing several constraints at once. Consider the constraint $[A \Rightarrow .7 \text{ and } B \Rightarrow .8]$! It is clear that ($A$ and B) must get quite a high probability here. (Look back to figure 1 to visualize this.) Even if (A or B) gets 1, so ($A - B$) must get 0.2, we will have $(0.7 - 0.2) = 0.5$ for the conjunction. If we give $1 - x$ to (A or B) then $A - B$ receives $0.2 - x$ and (A and B) gets $0.5 + x$, so we see that 0.5 is the *lowest* we can give it. But what value should we choose?

Another problem, also going beyond Jeffrey's rule, is to impose the constraint that the odds of A to B should become, say 2:1. In the diagram this means: set $(a + c) = 2(c + b)$. The question what the value c of the conjunction should be, is again wide open. In both of these problems, we are addressing an implicit partition of the possibilities into four sets: $(A\&B)$, $(A\&\bar{B})$, $(\bar{A}\&B)$, $(\bar{A}\&\bar{B})$. So as not to be too ambitious, I shall now focus our attention on one that really addresses only a three-fold partition, and still requires us to go beyond Jeffrey's Rule.

This is the Judy Benjamin problem (see van Fraassen 1981). It derives from the movie *Private Benjamin*, in which Goldie Hawn, playing the title character, joins the Army. She and her platoon, participating in war games on the side of the "Blue Army," are dropped in the wilderness, to scout the opposition ("Red Army"). They are soon lost. Leaving the movie script now, suppose the area to be divided in two halves, Blue and Red territory, while each territory is divided into Headquarters Company area and Second Company area. They were dropped more or less at the center, and therefore feel it is equally likely that they are now located in one area as in another. This gives us the following muddy Venn diagram, drawn as a map of the area:

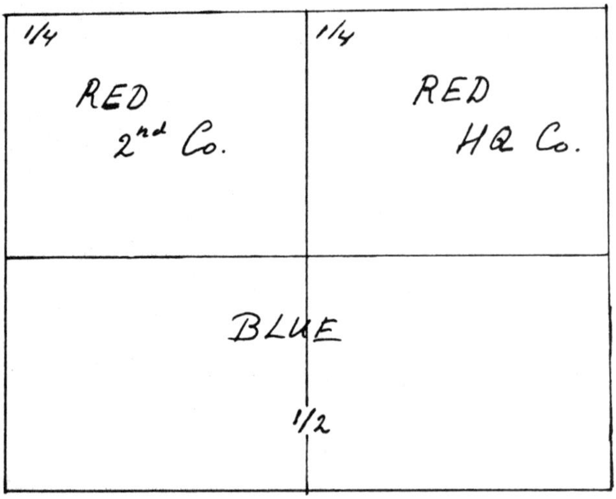

They have some difficulty contacting their own HQ by radio, but finally succeed and describe what they can see around them. After a while, the officer at HQ radios: "I can't be sure where you are. If you are in Red Territory, the odds are 3:1 that you are in HQ Co area . . . " At this point the radio gives out.

We must now consider how Judy Benjamin should adjust her opinions, if she accepts this radio message as the correct and sole constraint to impose. The question on which we should focus is: what does it do to the probability that they are in friendly Blue Territory? Does it increase, or decrease, or stay at its present level of ½? I originally proposed this question in a seminar at the University of Western Ontario, and I have canvassed not only that seminar, but several classes and conference audiences since, including the audience of the Pittsburgh Philosophy of Science Lecture Series. The "intuitive" or unreflective response has always been, overwhelmingly, that after the radio message about Red Territory, it should seem neither more nor less likely to Judy Benjamin that she is still in Blue.

Peter Williams brought out the main flaw in that sort of response. Suppose the radio officer had not said "3:1" but, in more general form:

> The probability that you are in HQ area, given that you are in Red Territory, equals q.

for some value q—¾ in our other example—which must lie between 0 and 1. Our task is to find a general rule to cover all. But if the radio officer had said 1—if q had equalled 1—he would have told her, in effect, "You are *not* in Red-2 Co area." In that case she could have used Simple Conditionalization, and wiped the mud of the top-left square in the diagram. This would have the result of giving Blue Territory ⅔ of the remaining mud. Hence the probality of Blue would go up; and we cannot make it a general rule that it stays the same.

XI. Approach Through Symmetry and Continuity

We have at this point already implicitly a general description of Judy Benjamin's problem, and several requirements on any solution. The first we have just stated:

1. If Simple Conditionalization is applicable, it should be applied.

This can be restated as follows: If $q = 1$ the odds of Blue to Red become equal to the prior odds of Blue to Red-HQ (while Red-2nd Co gets *zero*) and if $q = 0$ those odds become equal to the prior odds of Blue to Red 2nd Co (while Red-HQ receives *zero*).

There is a second requirement, even more obvious. Imagine that the voice on the radio had said "If you are in Red territory, you are equally likely to be in HQ as in 2nd Co area." Then he would simply have expressed the same opinion as Private Benjamin already had, so no change is called for:

2. If q equals the prior conditional probability (of HQ given Red), the probability of Blue (and hence all odds) should stay the same.

We have, so to say, a begin-point (new q = old q) and two end points (new q equals *zero* or *one*) and we know what *any* rule should do at those points.

We can immediately picture the simplest rule that satisfies these first two principles. Here is what it does in Judy Benjamin's case:

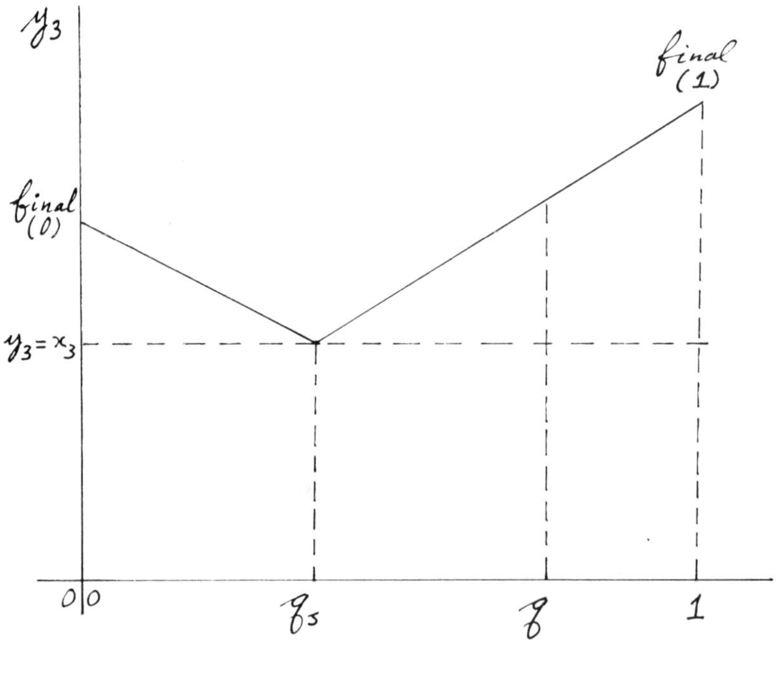

The solid line plots the new probability for Blue, against the new conditional probability q. This is the simplest because the end-points are connected with the begin-point by straight lines. Is this simplest rule the best? *No*, in fact it really is not any good. To see that we have to talk about symmetry.

I already explained, in connection with the justification of Jeffrey Conditionalization how a symmetry argument operates. Its crux is an assumption of what is relevant and what irrelevant. Or alternatively put, the crux is an account of which changes would affect the problem in some essential way. Now I'll propose two such "intuitions":

Intuition 1. The probability of Blue just "goes along for the ride"—it is not an essential factor in the solution.

Intuition 2. Whether a new conditional probability is posted for HQ given Red, or for not-HQ given Red (i.e. 2nd Co given Red) makes no real difference in the method of solution.

I mean the first to be the true core of the conviction that led to the first "intuitive solution" which most people go for—"Leave Blue alone!" The trouble was that the probability of Blue is not something independent of the other probabilities. If I change the problem by stating a new prior probability for Blue, I *have to* change the probabilities for Red HQ and Red 2nd Co as well because they must add up to *one*.

So we need a formulation in which the status of Blue is independent. This is given by the odds formulation: odds are in the same proportion as probabilities but need not sum to *one*.

Prior probab.: $\langle x_1, x_2, x_3 \rangle = \langle P(\text{Red 2nd Co}), P(\text{Red-HQ}), P(\text{Blue}) \rangle$

Prior odds $\langle 1, s, t \rangle$

where $s = x_2/x_1$ and $t = x_3/x_1$. Now the constraint says the odds of the second to the first must change from s:1 to some other number, say r:1.

Posterior odds $\langle 1, r, t' \rangle$

How does t change to t'? Intuition 1 tells us it must change in a fashion which is independent of the prior probability of Blue, i.e. independent of t:

$t' = \gamma t$, where γ is a function that does not depend on t

Of course the prior and posterior odds of Red HQ to Red 2nd Co *are* relevant, so we have as principle:

3. If the constraint is that the posterior odds must be of form $\langle 1, r, \ldots \rangle$ then there is a function $\gamma(s, r)$ such that prior odds $\langle 1, s, t \rangle$ are transformed into posterior odds $\langle 1, r, \gamma(s, r)t \rangle$

This was a very intuitive argument; the paper I referred to above, written with Hughes and Harman, gives a more rigorous version.

This result is really quite powerful. For example it eliminates the "straight-line rule" proposed at the beginning of this section.[3] In other words, if you use that straight-line rule, you will change the probability of Blue in a way that depends in a really essential way on what its prior probability was—it won't just go along for the ride.

The second intuition is also quite powerful. The radio-voice said "The conditional probability of HQ given Red equals ¾." If it had said "The conditional probability of 2nd Co given Red equals ¼," it would have meant the same thing. But now suppose it added "Sorry, that was a Spoonerism—wherever I said "HQ" I meant "2nd Co" and *vice versa*." Well, *that* makes no *real* difference. Calculations need not be done all over again—just "relabel" the areas as indicated. The way to think of it is this. By "relabelling" I mean interchanging "Red HQ" and "Red 2nd Co." Now suppose you have a prior opinion given, call it $\bar{x}$. First use your rule to find the correct posterior—call it $\bar{y}$. Next, go back to $\bar{x}$ and change it to another prior by relabelling—call it $\bar{x}'$. Use your rule *appropriately* to find a posterior for *that* prior—call this posterior $\bar{y}'$. Now change $\bar{y}'$ by relabelling. The result should be the same as the first time around: $\bar{y}'$ when relabelled should become $\bar{y}$. (Be sure however that when you applied your rule the second time, you also relabelled the constraint! That is why I said: use your rule appropriately.[4]) Again I won't go through the calculations here, but this implies something very pretty for the function mentioned in 3:

4. $\gamma(1/s, 1/r) = (s/r)\gamma(s, r)$

To see what it does and why it is important, let us ask: what is the simplest rule that satisfies 1-3? The answer is: γ always has value 1; the prior odds $\langle 1, s, t \rangle$ change simply to $\langle 1, r, t \rangle$. That rule is allowed by 1-3, but violates 4. It is a rule that "thinks" that the "direction" HQ-2nd Co bears on the solution to our problem—which it should not.

For completeness I must mention a fifth requirement, although it goes very little beyond the first one. When I stated 1 and 2, I was still speaking in terms of q (conditional probability of *HQ* given *Red*), whereas now we have switched to r (odds of *Red-HQ* to *Red-2nd Co*). It is easy enough to state part of 1, and 2, in odds terms.

1*. If $r = 0$ then $\gamma(s, r) = 1$
2*. If $r = s$ then $\gamma(s, r) = 1$

The second is easiest: if $r = s$ that means q is meant to stay the same (q = prior conditional probability) so the posterior probabilities are the *same* as the prior ones. For the first: if $r = 0$ that means that Red-HQ gets wiped out from the realm of possibility—i.e. we conditionalize on not-Red-HQ—i.e. $q = 0$. So the odds of Blue to Red become equal to the old odds of Blue to Red-2nd Co: $\langle 1, s, t \rangle$ becomes $\langle 1, 0, t \rangle$.

What is still left out is the case of $q = 1$. This means the odds of Red HQ to Red 2nd Co become 1:0 which is infinite. We can't work directly with infinity. We do know from the old version of 1 that $q = 1$ must be treated very much like $q = 0$. Indeed, the new probabilities in Judy Benjamin's problem must become $\langle 0, 1/3, 2/3 \rangle$. That is, the old odds $\langle 1, 1, 2 \rangle$ must be changed to $\langle 0, 1, 2 \rangle$. Or in general, $\langle 1, s, t \rangle$ must become $\langle 0, s, t \rangle$. So intuitively we need to say: if r becomes infinite, then $\langle 1, r, \gamma(s, r)t \rangle$ becomes the same as

$\langle 0, s, t \rangle$. The word "becomes" is metaphorical: it points to a picture of continuous development in time. So here is the modest proposal to capture the second part of 1 in odds terms, in as far as we can:

5. If $\gamma(s, r)$ is a continuous function of r then

$$\lim_{r\to\infty}[\gamma(s, r) - (r/s)] = 0$$

All the rules we shall seriously consider as solutions to Benjamin's problem will make γ continuous.

We now have a partial picture of how a rule determines in terms of r and s:

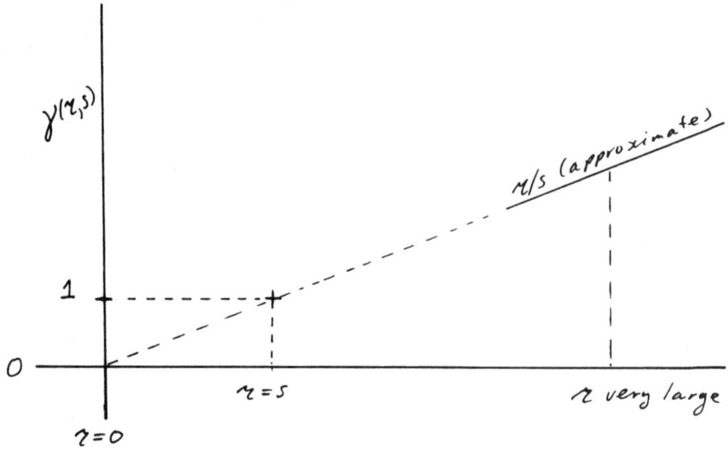

XII. Three Rules: INFOMIN, MTP and MUD

The requirements listed in the preceding section do not single out a unique rule. Nor would it be a good idea, I think, to look for the "simplest" or "nicest" rule that fits all of 1–4. For Judy Benjamin's problem is itself only the simplest of those which take us beyond Jeffrey Conditionalization. We need some more general view-point to lead us toward a more general solution. The image or metaphor that has recommended itself here is that of *nearness*: the person should always choose the posterior state of opinion, among those which satisfy the relevant constraints, which is nearest to his prior state. I do not know how one would justify this, or how it could issue in specific advice. Undoubtedly, some changes of opinion will look capricious, jumps to conclusions if not leaps of faith. But there are innumerable ways to spell out "nearness." Still, let us begin.

There exist three rules that give nice solutions to the Judy Benjamin problem. The first is the rule to minimize *relative information* (in the posterior with respect to the prior), the rule that I call INFOMIN. It was adapted from the information-theoretic approach in statistical mechanics, developed by Jaynes and others. The second is the rule of maximizing the *transition probability*, an analogue to quantum mechanics, conceived and developed by R.I.G. Hughes—we call it the MTP rule. The third rule, devised by Gilbert Harman, is the MUD rule. It is easily stated in terms of the muddy Venn diagram model: Private Benjamin should wipe away the minimum amount of mud so as to satisfy the constraint. A little reflection shows that to follow this rule, she should wipe mud off the Red-2nd Co area only if the new conditional probability is *higher* than the old one. In that case the odds 1:t of Red-2nd Co to Blue are changed by raising t. But those odds stay the same if she wipes mud off Red-HQ area—which is what she should do if the new conditional probability is lower. Of the three rules, this one is therefore so simple that it can (like Strict and Jeffrey Conditionalization) be explained without mathematics to speak of. Since we have discussed these rules in the abstract in recent joint papers (Hughes and van Fraassen 1985; van Fraassen, Hughes, and Harman 1986) I shall here immediately focus on the solutions they generate for the Judy Benjamin problem.

Let us first recall the symbols we are using. The *prior odds* are (x_1, x_2, x_3), equivalent to $(1, s, t)$. The constraint is that the new conditional probability of HQ given Red be q; or equivalently that the new odds of Red-HQ to Red-2nd Co be r—where $r = q/(1-q)$ and $q = r/(1+r)$. The solution is the description of the *posterior odds* as $(1, r, \gamma(s,r))$. Hence all a rule has to do, is to tell us what this factor $\gamma(s, r)$ is.

INFOMIN Solution: $\gamma(s, r) = (r/s)^q$

MTP Solution: $\gamma(s, r) = \left(\frac{1 + r}{1 + \sqrt{rs}}\right)^2$

MUD Solution: $\gamma(s, r)$ equals 1 if $r \leqslant s$ and equals (r/s) otherwise

All three make γ a continuous function of r and s. The five Principles 1–5 are all satisfied. This is a little difficult to see for I because $q = 0$ is equivalent to $r = \infty$, so let us leave that one till last. Principle 2 can be restated as: if $r = s$ then $\gamma(r, s) = 1$, which is obvious for all three solutions. Principle 3 is "built in" to the assertion that the posterior odds takes the form they do, expressible in terms of the factor γ, which is a function of r and s alone. The fourth Principle says that $\gamma(1/s, 1/r) = (s/r)\gamma(s, r)$, while Principle 5 says that $\gamma(s, r)$ approximates to (r/s) for large values of r. Both are checked by elementary calculations, recalling how r and q are related to each other. Thus we have only I left. Obviously $r = 0$ if $q = 0$, and $\gamma(s, 0) = 1$ in the case of each of the three rules. But if $q = 1$ then r is infinite. Now we see that Principle 5 covers this case, and it says that as q gets closer and closer to 1, the posterior odds begin to look like $\langle 1, r, (r/s)t \rangle$. Because r becomes so large, the number 1 becomes negligible, so those odds begin really to look like $\langle 0, r, (r/s)t \rangle$, which is the same as the odds $\langle 0, s, t \rangle$. And that effect-Simple Conditionalization on (not-Red 2nd Co)—is just what we wanted.

XIII. Comparison of the Rules

The first way to compare the three rules is to ask: by how much does the probability of Blue go up for Judy Benjamin? Let us ask this for an *upward* revision of the conditional probability of HQ given Red—i.e. in our symbolism, $r > s$. Then we find:

for all values of $r \geqslant s$: $\gamma(\text{MTP}) \leqslant \gamma(\text{INF}) \leqslant \gamma(\text{MUD})$

where γ(———) is the factor as specified by rule ———.[3] If you look at what happens when r is increased, you see that γ increases too—in the case of MUD of course by the constant factor $1/s$, hence a straight line if you plot against r. The other curves are not so simple—but we know from Principle 5 that they must "eventually" coincide with the one for MUD. Since they coincide also at $r = s$, when $\gamma = 1$, the other two are "concave," it looks like this:

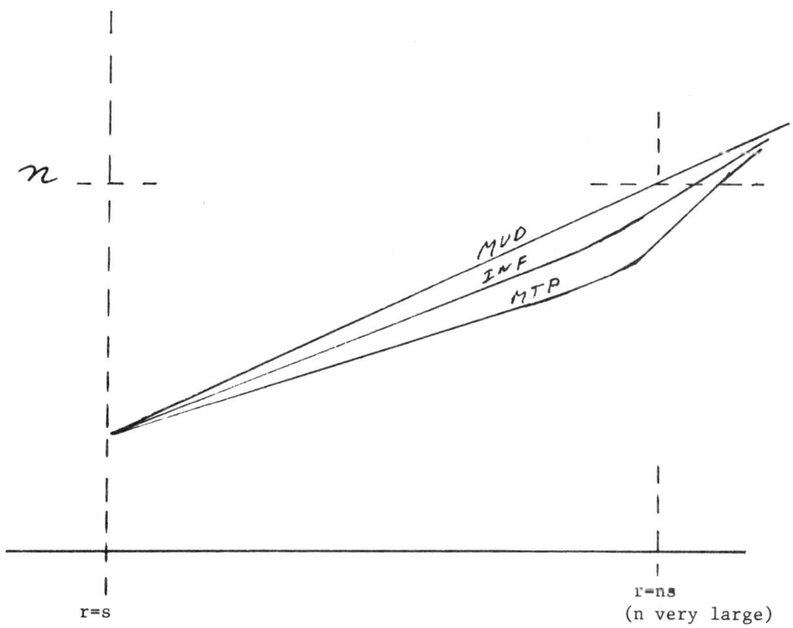

So when the revised conditional probability comes close to 1—i.e. r very large—the probability of Blue is still going up at the same rate according to MUD, but going up faster than it was according to INF, and still faster according to MTP. Earlier on, however, with the revised conditional probability closer still to the old ½, MTP prescribed the smallest increase in the probability of Blue and MUD the largest. If you remember this below you won't be to surprised at an effect that I shall describe as the "wavering effect." It will be important in the evaluation of these rules.

If you look at the probabilities instead of the odds, and think about increasing revised conditional probabilities for HQ given Red, you'll see a curious feature. The probability of 2nd Co Red goes steadily down toward zero, and the probability of Blue goes steadily up toward ⅔. But the probability of Red HQ goes up to a maximum, and then goes down (toward ⅓). The approximate maximum values are:

INFOMIN: maximum ca. 0.36887 at $q = 0.877$
MTP: maximum ca. 0.40064 at $q = 0.916$

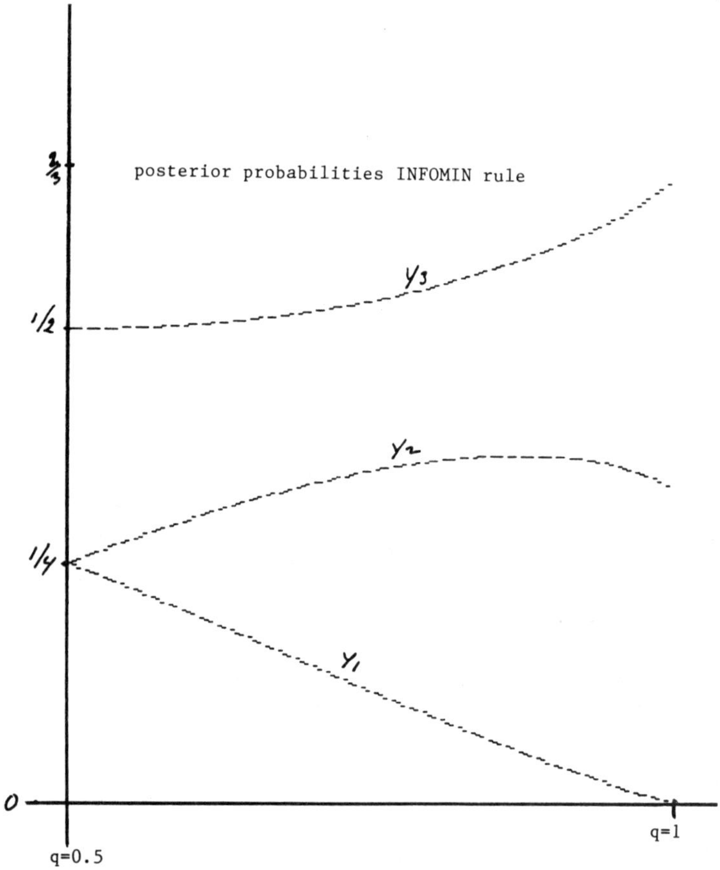

For MUD the situation is different. Let us consider $\gamma(s, r)$ for $r > s$, i.e. $\gamma(s, r) = r/s$. The posterior odds are $\langle 1, r, t(r/s) \rangle$ hence the posterior probabilities $\langle y_1, y_2, y_3 \rangle$ are $\langle 1/f(r), r/f(r), tu/sf(r) \rangle$ where

$f(r) = 1 + r + t(r/s)$. If we differentiate there posterior probabilities with respect to r we find

$$dy_1/dr = (-1 - t/s)/f(r)^2 < 0$$

$$dy_2/dr = 1/f(r)^2 > 0 \text{ but decreasing with } r$$

$$dy_3/dr = (t/s)dy_2/dr > 0 \text{ but decreasing with } r$$

Hence we conclude that the curves for y_1, y_2, y_3 have no maximum or minimum between the beginning and ending points. The probability for Red-2nd Co decreases, and the other two increase, for as long as r increases—however, the rates of increase and decrease level off for high values of r.

XIV. The Horse Race Continued: Quasi-Empirical Criteria

If we have no general grounds to prefer either solution to the other, I thought at one point, it might still be that a person using the one rule could have an advantage over those who use the others. One feasible thought experiment is like the one we used to "test" Jeffrey Conditionalization. Let us use the same three horses, and let our agents witness daily run-offs in which two of the three horses compete against each other. Each day D_n the onlooker may impose new odds for one horse H_n^1 against another H_n^2, and we assume the imposed odds are actually c_n^1:c_n^2 in agreement with the "real" chances of winning. Will their opinions converge to this real chance function, and if so, how fast? Well, we ran the race (on an IBM PC), and (fortunately or unfortunately) the INFOMIN and MTP rules did quite well—more or less equally well! Display No. 4 gives the results for the same priors and targets we listed in section VIII, in the same order.

The interesting fact, jumping to the eye as it were, is that this list is again in the "right" order: the needed number of steps increases again (except in the perhaps negligible first column) as you go down the list. We have no explanation for this.

But then the race was run again with a third bettor, who used the MUD rule. And *he* beat the others hands down. Display No. 5

Display No. 4

	Prior	Target		Degree of Approximation			
				1/10	$1/10^4$	$1/10^8$	$1/10^9$
(A)	1,1,1	$1,10^3,10^6$	*INF*	3	3	3	6
			MTP	3	3	3	6
(B)	1,1,1	1,10,100	*INF*	3	6	12	13
			MTP	3	6	12	15
(C)	1,1,1	1,1,20	*INF*	3	8	15	17
			MTP	3	9	15	18
(D)	3,3,4	1,2,7	*INF*	6	9	20	24
			MTP	6	9	21	23
(E)	7,2,1	1,2,7	*INF*	6	11	21	24
			MTP	6	12	21	24
(F)	5,6,7	6,7,8	*INF*	1	8	21	24
			MTP	1	8	21	24
(G)	5,6,7	7,8,9	*INF*	2	9	22	25
			MTP	2	9	22	25
(H)	1,5,25	1,2,4	*INF*	6	12	23	27
			MTP	6	12	23	26

shows some of the results—extending the table would make MUD look even better.

Display No. 5

Prior	Target	0.1:	INF	MTP	MUD	0.0001:	INF	MTP	MUD
1,1,1	$1,10^3,10^6$		3	3	3		3	3	3
1,1,1	1,10,100		3	3	3		6	6	3
3,3,4	1,2,7		6	6	3		9	9	3
7,2,1	1,2,7		6	6	3		11	12	3
5,6,7	7,8,9		2	2	1		9	9	3
1,5,25	1,2,4		6	6	2		9	9	3

Wonderful! Has the simplest and most visualizable rule turned out to be also the best?

Unfortunately, no; for there is another criterion that applies to less ideal learning situations. In our thought experiment about the horse race so far, the bettors were ideal perceivers: each time they saw two horses run against each others they espied the "real" conditional target probabilities. But there is one feature quite noticeable in each rule, that may make for difficulty in other cases. Each rule, applied to the Judy Benjamin case, makes the probability of Blue go up at each application. Let us ask this embarrassing question: if a bettor, watching the same pair of horses race against each other several times, wavered back and forth between two conditional probabilities q_1 and q_2 (for the first horse winning if either does), how long before he is 99% certain that neither will win? Or, equivalently, if Judy Benjamin wavers back and forth between q_1 and q_2, *how long before she is 99% certain she is in the Blue Area anyway*? It was easy to simulate this situation on a PC also. Display No. 6 gives are some results:

Display No. 6

Prior	q_1	$q_2 = (1 - q_1)$	INF	MTP	MUD
1,1,2	⅔		22	41	9
	¾		11	18	6
1,2,3	⅔		22	42	9
	¾		11	18	7
2,1,3	⅔		21	41	8
	¾		10	18	6
5,6,7	⅔		24	45	9

Now the fewer the steps, the worse the performance of the rule. So MUD does very badly—it gives 0.99 probability to the "third" possibility in less than 10 wavering steps. MTP does best, allowing the waverer some grace, while INF is somewhere in between, though closer to MUD than to MTP.

XV. The Direction of General Probability Kinematics

The general problem of probability kinematics is this: given a prior state of opinion, and some constraint on the posterior opinion, what is the rule for forming the posterior state of opinion? So stated there is a presupposition, namely that there is a "right" posterior, determined uniquely by prior and imposed constraint. I do not think this presupposition should be accepted as a principle. However, we can put the question without prejudice as follows: *If* all I want to do is transform my present opinion, so as to satisfy a given constraint, without taking into account any other factors, how shall I re-form my opinion? And I think that this must be the relevant question most of the time, when we respond to experience, and do not take time to deliberate about theoretical alternatives and innovations.

Qualifications could to some extent be honored without changing the form of the problem. Suppose that an extra relevant factor is certain presuppositions, assumptions, sets of alternatives or whatever—features of my present thinking that are to be kept constant. These could be added into the constraint, so as to make it more stringent.

Of course there can be no pretense that we have come anywhere near solving the general problem here. If we want to go on piecemeal, here are three special instances of the general problem, which we have not yet covered:

Impose constraints of form:

(a) change simutanousely both the probabilities of A and of B, to 0.3 and 0.7 respectively,
(b) change the odds of A to B, to 3:1
(c) change the expectation values of quantities X, Y, Z, . . . , to 4.1, 4.2, 4.3, . . . respectively

There are special cases of (a) and (b) that we have handled. For example, if B is just (not-A) then (a) is simply Jeffrey Conditionalization. And if A and B are disjoint, then (b) is the Judy Benjamin problem. But if A and B can be any two propositions, then our discussion has not covered (a) and (b). Problem (c) is the most general we have touched on yet. Indeed, (a) and (b) are special cases of (c). So a rule that handles (c) handles all the problems we have come across so far.

Does this mean we should simply concentrate on (c), i.e. Expectation constraints? There is a good deal of literature on this in connection with INFOMIN, which is indeed applicable to such constraints. The physicist Edwin T. Jaynes proposed this rule in statistical mechanics, and had considerable success with it (see Levine and Tribus). The rule has also had successful engineering applications in data analysis, specifically photographic image enhancement (Skilling). Lately there have appeared new deductions of the rule from much weaker premises, including interesting symmetry arguments (Shore and Johnson; Tikochinsky et al.). All of this suggests an affirmative answer to our question.

But the prominence of INFOMIN is due in great part to its being the only rule *that is known and has been investigated at length*, which can handle expectation value constraints. Mathematically speaking, other such rules exist (see Diaconis and Zabell), but we don't know them in the same way. The symmetry arguments mentioned above should rule out these others from our consideration, but symmetry arguments are often based in part on desiderata or assumptions which are not totally incontrovertible.

Apart from that, should we rule out from consideration, say, any rule that solves the Judy Benjamin problem, but cannot be extended to Expectation constraints? The general principle behind such a decision—*require such rules to be extendible to other sorts of constraints*—would make us discard INFOMIN as well. For it cannot be used to impose such a constraint as, for example,

(d) change your probabilities so as to make A and B statistically independent

i.e. so that the posterior P' has $P'(A \text{ and } B) = P'(A).\ P'(B)$.

If the Judy Benjamin problem has no rationally compelled *unique* answer, then—mathematically and logically speaking—it can still be a special instance of a more general problem, which does have a unique solution. (After all, solutions take the form "here is a rule that covers all the cases in your target class.") But that does not mean automatically that the Judy Benjamin problem has a unique solution. And for someone who asks: "am I rationally compelled to change my opinions in such and such a fashion?" it is not a complete answer to be told "If you decide to adopt a rule for solving all such problems with constraints definable in terms of posterior expectation values, then *yes*."

Yet, even if incomplete, this would be a very helpful answer.

Hence our very next task should be to examine those "deductions" of the INFOMIN rule and see how compelling they are. This can be the first entry in our next section.

XVI. Open Problems in the Alternative Epistemology

Richard Jeffrey seems to me to be correct in regarding this approach as the outcome of a long development, that has always posed an alternative in epistemology. It is at the same time a young field, with many problems remaining. I will list a few, keyed to the sections above.

(II) What exactly is a *judgement*? In "Belief and the Will" I argued that a judgement is definitely not an autobiographical statement describing one's own state of mind, but the expression of something like a commitment or intention. Commentators have not agreed.

(V) Could there be, in this alternative tradition, anything like induction, abduction, or more generally, reasons for belief? In the strict Bayesian model, you should already have conditionalized on the evidence when you begin to think about it—and no more needs or should be done. Any rule you could add would either follow from or conflict with the rule of Simple Conditionalization—hence be trivial or *de trop*. In "Belief and the Will" again, I have argued that some reconciliation between the two traditions is possible. In the "voluntarist" model of experience I offered in section V, there is some leeway: the acceptance of a constraint and possibly deliberation about *how* it should be satisfied, intervenes between experience and the "creation" of one's posterior state of opinion.

(VIII) Answering Levi's objection, I took recourse to a "practical" demonstration that "learning" from or by experience is possible via Jeffrey Conditionalization. This needs some theortical underpinning. First of all, I would like a theorem that said that the sequence of probability functions produced by the process must always converge to the target. Secondly, I'd like this theorem, if possible, to be of the nice sort that shows *how* the difference from the target diminishes in the long run. (In the Appendix by John Collins we now have a proof of the theorem; it is not clear yet whether this gives information about rates of convergence.) In addition, we want a rule for the *joint* imposition of such constraints. The general INFOMIN rule will do this; but of course we also need some idea of what is a *good* way to do it.

(IX) Are there more requirements besides 1–5 for the Judy Benjamin solution? Or do we really stand here before a problem without a unique solution? How shall we weigh the two quasi-empirical tests? Can 1–5 be generalized to other sorts of problems?

(X) Diaconis and Zabell have discussed other rules that could be used, mainly ones that look like INFOMIN in some way; could we study *families* of rules? In the PSA paper with R.I.G. Hughes, we discussed the general rationales for INF and MTP.

(XI) As for section (VIII) we must ask whether the apparent convergence to the target is provably there, in general or within some range; also, what indicators there might be of how fast the approach would go.

Finally, looking at just that "indicator" problem, I found another one. Hughes' MTP rule uses a familiar geometric distance measure (the "transition probability" of quantum mechanics). Relative information is not a distance measure, because it is not symmetric. But we can define a symmetric measure, by taking the average of the relative information in both directions. It turns out (quasi-empirically) that this agrees in nearness ordering (of, for example, our prior-target pairs (A)—(H)) with the nearness-ordering induced by the MTP distance measure. We have not yet found the mathematical reason for this.

There are therefore a number of interesting open problems, both philosphical and technical.[5]

NOTES

1. For a better overview of general epistemology as I see it, and probability's place therein, I refer to my "Empiricism in the Philosophy of Science" and "Belief and the Will."

2. These issues are discussed to some extent in Jeffrey's early article "Probable Knowledge." Recalling it raises the question whether we can

help what our primary response will be like. I am not at all sure one way or the other, even if most of these responses, like most simple actions, are made "automatically."

3. To see this, it suffices to calculate $\gamma(s, r)$ for various cases in which s and r are the same, while t is different. If the prior is $\langle x_1, x_2, x_3\rangle$ and the posterior is $\langle y_1, y_2, y_3\rangle$ then $s = x_2/x_1$, $r = y_2/y_1$, and $\gamma(s, r) = (y_3/y_1) \div (x_3/x_1)$. So now all we need to know is how to calculate the posterior by that "straight-line rule" in general. For Judy Benjamin's prior $\langle 1/4, 1/4, 1/2\rangle$ and $q = 3/4$ which is half way between the old conditional probability ½ and the maximum 1, it tells us to set the new probability of Blue half-way between its prior value and its maximum which is $x_3/(x_3 + x_2 + 0)$. Here are some results (with appropriate values displayed):

Prior	Posterior	y_3/y_1	$(y_3/y_1) \div (x_3/x_1)$
¼, ¼, ½	5/48, 15/48, 28/48	5.6	2.8
.2, .2, .6	0.08125, 0.24375, 0.675	8.31	2.77
.1, .1, .8	0.0389, 0.1167, 0.8444	21.7	2.7125
.05, .05, .9	0.01908, 0.0572, 0.923684	48.414	2.689

Thus, although r and s are the same in all the examples, the ratio x_3/x_1, does not get changed by a constant factor. Hence 3 is violated; otherwise the last column would have shown the same number throughout.

4. Here is the calculation that yields 4. Prior $\langle 1, s, t\rangle$ with constraint that the new odds should be r:1, yields first posterior $\langle 1, r, \gamma(s, r)t\rangle$. The relabelled prior is $\langle s, 1, t\rangle = \langle 1, 1/s, t/s\rangle$. Apply the rule with the constraint relabelled, i.e. the new odds are 1:r which is $(1/r)$:1, and get new posterior $\langle 1, 1/r, \gamma(1/s, 1/r)(t/s)\rangle$. Relabel again: this becomes $\langle 1/r, 1, \gamma(1/s, 1/r)(t/s)\rangle = \langle 1, r, \gamma(1/s, 1/r)(t/s)r\rangle$. This second posterior must equal the first; hence 4.

5. Support for this research by the National Science Foundation is gratefully acknowledged. I want to thank Gilbert Harman and R.I.G. Hughes for their comments on an early version (presented in Princeton, April 1985, as "Opinion: Reflections on Richard Jeffrey in his Absence"); parts of this paper form a less technical exposition of the results reported in our forthcoming joint paper.

BIBLIOGRAPHY

Diaconis, P. and S. Zabell, "Updating Subjective Probability," *Journal of the American Statistical Association* vol. 77 (1982), pp. 822–830.

Field, M., "A Note on Jeffrey Conditionalization," *Philosophy of Science*, vol. 45 (1978), pp. 361–367.

Hacking, I., "Slightly More Realistic Personal Probability," *Philosophy of Science*, vol. 34 (1967), pp. 311–325.

Hughes, R.I.G. and B.C. van Fraassen, "Symmetry Arguments in Probability Kinematics," *Philosophy of Science Association 1984*, vol. 2., forthcoming, 1986.

Jeffrey, R., "Bayesianism with a Human Face," *Minnesota Studies in the Philosophy of Science*, vol. X (1984), pp. 133–156.

_____ "Judgmental Probability and Objective Change," *Erkenntnis*, vol. 24 (1986), pp. 5–16.

_____ *The Logic of Decision* (Chicago: University of Chicago Press, 1983).

_____"Probable Knowledge", in I. Lakatos (ed.) *The Problem of Inductive Logic* (Amsterdam: North-Holland, 1968), pp. 166–180.

Levi, I. "Probability Kinematics," *British Journal for the Philosophy of Science*, vol. 18 (1967), pp. 197–209.

Levine, R.D. and M. Tribus, *The Maximum Entropy Formalism* (Cambridge, Mass.: MIT Press, 1979).

Shafer, G., "A Subjective Approach to Conditional Probability," *The Journal of Philosophy*, vol. 12 (1983), pp. 453–466.

Shore, J.E. and R.W. Johnson, "Axiomatic Derivation of the Principle of Maximum Cross-Entropy," *IEEE Trans. Information Theory*, IT-26 (1980), pp. 26–37.

Skilling, J. "The Maximum Entropy Method," *Nature* 309 (June 28, 1984), pp. 748–749.

Teller, P. "Conditionalization, observation, and change of preference," in W.L. Harper and C.A. Hooker (eds.) *Foundation of Probability, Statistical Inference and Statistical Theories of Science* (Dordreht: Reidel, 1976). p. 205.

Teller, P. and A. Fine "A Characterization of Conditional Probability," *Mathematics Magazine*, vol. 48 (1975), pp. 267–270.

Tikochinsky, Y; N.Z. Tishby, and R.D. Levine, "Consistent Inference of Probabilities for Reproducible Experiments," *Physical Review Letters* vol. 52 (1984), pp. 1357–1360.

van Fraassen, B.C. "Belief and the Will," *The Journal of Philosophy*, vol. 81 (1984), pp. 235–256.

_____ "A Demonstration of the Jeffrey Conditionalization Rule," *Erkenntnis*, vol. 24 (1986), pp. 17–24.

_____ "Empiricism in the Philosophy of Science," in P.M. Churchland and C.A. Hooker (eds.) *Images of Science* (Chicago: University of Chicago Press, 1985), pp. 245–308.

_____ "A Problem for Relative Information Minimizers," *British*

Journal for the Philosophy of Science, vol. 32 (1981), pp. 375–379.

______ "Rational Belief and Probability Kinematics," *Philosophy of Science*, vol. 47 (1980), pp. 165–187.

______ "Shafer on Conditional Probability," *Journal of Philosophical Logic* vol. 12 (1983), pp. 467–470.

van Fraassen, B.C.; R.I.G. Hughes, and G. Harman "A Problem for Relative Information Minimizers, Continued," *British Journal for the Philosophy of Science*, (1986) (forthcoming).

APPENDIX: JEFFREY CONDITIONALIZATION—A CONVERGENCE PROOF

J.D. COLLINS

One of the open problems in Bas van Fraassen's paper "Symmetries of Personal Probability Kinematics" concerns convergence in the Jeffrey Conditionalization "horse race" example. The following is a proof that the probability functions converge to the targets.

Let the possible outcomes of the race be $H_1, \ldots ,H_n$ with objective chances $P_1, \ldots P_n$. An arbitrary initial subjective probability assignment to $H_1, \ldots H_n$ is revised in a sequence of steps. At the ith step ($i \leqslant n$), the subjective probability for H_i is set at the objectively correct P_i, while all the other values obtained at the preceding stage are adjusted according to the Jeffrey Conditionalization rule. After the nth step, we return to the first and repeat the procedure. At each stage of this process the correct assignment from the previous stage is upset; nevertheless, the resulting sequence converges to the target probabilities $P_1, \ldots P_n$.

We consider the case $n = 3$. The general result can be proved in a similar way. Let the initial probabilities be a_0, b_0, c_0 and the target probabilities a, b, c, where $a_0 + b_0 + c_0 = a + b + c = 1$. We assume $0 < a, b, c < 1$, as the other cases are trivial.

The Jeffrey rule yields a recursive definition of three sequences a_i, b_i, c_i:

$$a_i = \begin{cases} a_{i-1}\left(\dfrac{a+b}{a_{i-1}+b}\right) & \text{if } i = 0 \bmod 3 \quad (1) \\ a & \text{if } i = 1 \bmod 3 \quad (2) \\ a\left(\dfrac{a+c}{a+c_{i-1}}\right) & \text{if } i = 2 \bmod 3 \quad (3) \end{cases}$$

$$b_i = \begin{cases} b\left(\dfrac{a+b}{a_{i-1}+b}\right) & \text{if } i = 0 \bmod 3 \quad (4) \\ b_{i-1}\left(\dfrac{b+c}{b_{i-1}+c}\right) & \text{if } i = 1 \bmod 3 \quad (5) \\ b & \text{if } i = 2 \bmod 3 \quad (6) \end{cases}$$

$$c_i = \begin{cases} c & \text{if } i = 0 \bmod 3 \quad (7) \\ c\left(\dfrac{b+c}{b_{i-1}+c}\right) & \text{if } i = 1 \bmod 3 \quad (8) \\ c_{i-1}\left(\dfrac{a+c}{a+c_{i-1}}\right) & \text{if } i = 2 \bmod 3 \quad (9) \end{cases}$$

The problem is to show that $a_i \to a$, $b_i \to$ b and $c_i \to c$ as $i \to \infty$.

By (2), $a_1 = a$. If $b_1 = b$ and $c_1 = c$ then we are done, so we may assume that either $b_1 > b$ or $c_1 > c$. Suppose the former is true (the proof is similar in the other case). Then $b_1 > b$ and $c_1 < c$.

By (3), $a_2 = a(a + c)/(a + c_1) > a_1 = a$ and by (6), $b_2 = b$.

So it must be the case that $c_2 < c$, since $a_2 + b_2 + c_2 = a + b + c = 1$.

But by (9), $c_2 = c_1(a + c)/(a + c_1) > c_1$, so in fact, $c_1 < c_2 < c$.

The above illustrates the proof of the following inequalities, true for all k:

$$a_{6k+5} < a_{6k+6} < a < a_{6k+3} < a_{6k+2}$$
$$b_{6k+3} < b_{6k+4} < b < b_{6k+1} < b_{6k}$$
$$c_{6k+1} < c_{6k+2} < c < c_{6k+5} < c_{6k+4}$$

These inequalities show that we need only consider the subsequences:

(i) $a_2, a_8, \ldots, a_{6k+2}, \ldots$ (iv) $a_5, a_{11}, \ldots, a_{6k+5}, \ldots$

(ii) $b_6, b_{12}, \ldots, b_{6k}, \ldots$ (v) $b_3, b_9, \ldots, b_{6k+3}, \ldots$

(iii) $c_4, c_{10}, \ldots, c_{6k+4}, \ldots$ (vi) $c_1, c_7, \ldots, c_{6k+1}, \ldots$

Each of the subsequences (i), (ii), (iii) is bounded below, and each of (iv), (v), (vi) is bounded above, by the appropriate target probability. Hence to prove the whole result it suffices to show that the first three subsequences are monotonic decreasing and the other three monotonic increasing. Consider subsequence (i). Repeated application of (3), (8) and (4), in that order, yields:

$$a_{6k+8} = \left(\frac{a \quad a+c}{a + \left(\frac{c \quad c+b}{c+b\left(\frac{b+a}{b+a\left(\frac{a+c}{a+c\left(\frac{c+b}{c+b\left(\frac{b+a}{b+a_{6k+2}}\right)}\right)}\right)}\right)}\right)}\right)$$

Inspecting this equation we see that to show $a_{6k+8} < a_{6k+2}$ it is enough to show that if $w > 1$ and $0 < x, y, z < 1$, then

$$1 < w' = \frac{x + y}{x + y\left(\dfrac{y + z}{y + zw}\right)} < w.$$

Now in fact,

$$w - w' = \frac{(xy + xzw + y^2)(w - 1)}{(xy + xzw + y^2 + yz)} > 0,$$

$$\text{and } w' - 1 = \frac{yz(w - 1)}{(xy + xzw + y^2 + yz)} > 0.$$

The other subsequences are shown to be monotonic in the same way.

COHERENCE

BRIAN SKYRMS

I. Pumping gold

A very rich and reliable bookie posts odds on a horse race. He values a contract which pays a pound of gold if Stewball wins at one ounce of gold. He values a contract which pays a pound of gold if Molly wins at one ounce of gold. He will buy or sell these contracts in any quantity at what he considers the fair price or better. Suppose that he also deals in contracts which pay a pound of gold if either Stewball or Molly win, and that he values these at four ounces of gold. If you had two ounces of gold, you could then buy separate contracts for Stewball and Molly for a net outlay of two ounces of gold and sell back a disjunctive contract on Stewball or Molly winning for four ounces of gold. With these you can now buy two seperate contracts each on Stewball and Molly, and sell back two disjunctive contracts, etc. If you don't have two ounces of gold, perhaps your banker will lend them to you on the strength of your prospects. You are able to make a *Dutch Book* against the bookie—a finite number of bets whose net gain is positive no matter what the outcome of the race—because his evaluation of the disjunctive contract does not cohere with his evaluation of the seperate contracts on its disjuncts.

Coherence requires that if the bookie values the seperate contracts at one ounce of gold each, he values the disjunctive contract which pays off exactly as holding both seperate contracts together would, at two ounces of gold. Coherence also requires that a contract that pays a pound of gold no matter what is worth a pound of gold; and that a contract that pays off a pound of gold in some circumstances and requires no payment otherwise, has non negative value (i.e. the bookie would not pay someone to take it off his hands.) What are we to make of these obvious, but nevertheless remarkable, facts?

deFinetti (1937) used them as a foundation for the theory of personal probability. The value to you (in pounds of gold) of a contract which pays one pound of gold if *p*; nothing otherwise is taken as your personal probability of *p*. Then if your personal probabilities are *coherent*—i.e. a cunning bettor cannot make a dutch book against you in a finite number of bets-your personal probabilities must satisfy the mathematical laws:

(1) Probability is non-negative
(2) Probability of a tautology is one.
(3) If p;q are mutually exclusive alternatives, the probability of their disjunction is the sum of their probabilities.

The mathematics of these classical dutch book arguments is rather trivial and is certainly well understood, but their epistemological status remains a matter of philosophical controversy. And the matter of how one is to change from one coherent probability assignment to another under the pressure of new evidence is one which is not settled by the classical static coherence arguments. In this paper I would like to do two things: (I) suggest that when properly viewed the classical coherence arguments have more epistemological force than many critics have been willing to grant them, and (II) report some recent work on *dynamic* coherence.

II. Preaching to the Converted

The classical result is that a bookie whose posted betting ratios do not conform to the probability calculus can have a dutch book made against him. There are two approaches one can take in analyzing the significance of this result. One can (A) focus on the pragmatics of posting odds and attempt to construct an appropriate general theory or (B) one can try to locate the reason for the dutch book in a deeper inconsistency.

If one pursues course (A) it rapidly becomes clear that a general theory of optimal betting behavior will be extremely complicated and will lack the simple connection with degree of belief postulated in the classical dutch book situation. The bookie may know something about the bettor or bettors he is likely to encounter. He may know something about their propensities to bet when offered various options. Depending on the details he may be best advised to posted betting ratios which diverge widely from his true degrees of belief, and in certain situations would do well by posting *incoherent* betting ratios. To take an extreme and absurd case you can imagine that you will only face one bettor and that you know that when confronted with incoherence he will—for whatever reason—run the dutch book in reverse, pumping gold into your treasury instead of out. There are less trivial cases, however, whose analysis is of interest [e.g. see Adams and Rosenkrantz (1980)].

Ramsey [1931] takes the second approach:

> If anyone's mental condition violated these laws, his choice would depend on the precise form in which the option were offered him, which would be absurd. He could then have book made against him by a cunning bettor and would stand to lose in any event.

For Ramsey, the cunning bettor is a dramatic device and the possibility of a dutch book a striking symptom of a deeper incoherence. He suggests that coherence is just what logicians call "extensionality." If we are interested in the status of the dutch book theorems as a foundation for the theory of personal probability, this direction is clearly the one which merits investigation.

What will not do, is to confuse the two sorts of inquiry and to attempt to use the spectre of the cunning bettor as a bogie to coerce compliance with the laws of probability. Perhaps one should not let the bettor choose which side of the bet to take [Baille (1973)]. Perhaps one is not likely to run across a cunning enough bettor at all [Putnam (1975), Kyburg (1978), Chihara and Kennedy (1979), Glymour (1980)]. Must a rational person always act as if he is confronted by a cunning bettor, ever when he isn't [Jackson and Pargetter (1976)]? If one is interested in the status of the theory of personal probability, these concerns operate at the level of the dramatic device, rather than that of the underlying logic.

Let us look a little more closely at the structure of the dutch book argument for additivity, keeping in mind all the while the Ramsey point of view. Assume for the time being that all that is of value in the betting situation is payoffs in gold, and that gold has constant marginal utility (i.e. an ounce of gold counts for as much to the bookie whether added to big winnings or big losses). A *betting arrangement*, B, is a function from possible states of affairs to payoffs in gold. A *bet* on a proposition, p, is a betting arrangement which returns some gain, a, whever p, is true and some loss, b, whenever p is false. The *aggregate*, $B^1 \# B^2$, of two betting arrangements is the betting arrangement which has at each possible state of affairs, w, the sum of the payoffs at w of the constituent bets, B^1; B^2:

$$B^1 \# B^2 (w) = B^1 (w) \# B_2 (w) \qquad \text{for all } w.$$

Aggregation corresponds to the net effect of holding several bets. If the bookie is to be consistent, his prices should be additive over aggregation:

$$V (B^1 \# B^2) = V (B^1) + V (B^2)$$

since if he sells you B^1 for an ounce of gold and B^2 for two ounces of gold, he has *in effect* sold you $B^1 \# B^2$ for three ounces of gold. But then it is simply a matter of logic that if p;q are incompatible propositions, the aggregate of a bet that returns one pound of gold if p; nothing otherwise and a bet that returns one pound of gold if q nothing otherwise *is* a bet on the proposition p *or* q which returns one pound of gold if p or q; nothing otherwise. So, by definition, probabilities of incompatible alternatives add. [As an aside I might add that the argument works just as well for countable as for finite additivity. See Adams (1962).]

Some critics [e.g. Kyburg (1978), Schick (forthcoming)] have argued that the dutch book argument falls short essentially because additivity over aggregation fails in many real world situations. Non-monetary goods exhibit complementarities. A bet which returns a sow and a boar if the coin comes up heads may be worth more to you than the sum of the values of the separate bets. And monetary goods may have declining marginal utility. A bet which returns 100 pounds of gold if p may not have for you 100 times the value of a bet which returns one pound of gold if p. These phenomonae have been remarked upon at least since Daniel Bernoulli, and Ramsey and deFinetti were well aware of them. To discuss them with any clarity, we need a theory of utility.

von Neumann and Morgenstern (1944) rediscovering an idea of Ramsey (1931) suggested that we measure utility roughly as follows. We take the least desirable payoff in the decision situation and conventionally choose it as the zero point of our utility scale; we take the most desirable payoff and conventionally give it utility of one. Then using some chance device with known chances, and whose outcomes are "ethically neutral" we judge another possible payoff to have value x if we are indifferent between it and a wager which gives a chance x of getting the most desirable payoff, otherwise the least desirable payoff. [Ramsey goes further, dispensing with the known objective chances in favor of subjective surrogates, but there is no time here to go into these details.] In terms of such a theory of utility, complimentary goods and declining marginal utility of monetary instruments makes perfect sense. What then do these phenomonae show about additivity over aggregation?

Let us say that the *physical* aggregate of two bets, B^1; B^2 pays off in each possible situation both the physical goods that B^1 pays off and the physical goods that B^2 pays off. Let us say that B^3 is the *mathematical* aggregate of B_1 and B_2 if B_3 is such that at each possible situation, the utility of its payoff is equal to the sum of the utilities of the payoffs of B^1 and B^2. The payoffs are in goods which exhibit complimentarities or variable marginal utilities, a physical aggregate of two bets may not be a mathematical aggregate of them. Such cases provide examples of failures of additivity of utility over physical aggregation, but not failure of additivity over mathematical aggregation. To say that such examples provide counter examples to the additivity of value over aggregation is, I shall argue, a little like saying that the fact that adding a quart of water to a quart of alcohol does not yield two quarts of fluid provides a counter example to the laws of arithmetic.

In the general setting in which we are now working, the mathematical aggregate of two bets need not represent the net payoff resulting from undertaking both B^1 and B^2. What is the point of mathematical aggregation? It is part of a descriptive language. Suppose that propositions *p;q* are incompatible and that one has a betting arrangement that has a payoff of utility *x* if either *p* or *q* is true and a payoff of utility zero otherwise. Such a betting arrangement can be correctly described either as a bet on the disjunctive proposition, *p or q* or as the mathematical aggregate of two bets, one of which yields a payoff of utility *x* if *p* is true, utility zero otherwise; the other of which yeilds a payoff of utility *x* if *q* is true, utility zero otherwise. It is additivity over *mathematical* aggregation, not physical aggregation, that is important for the dutch book argument.

Now we have abstracted away from the inessential features of bookmaking so that bets and betting arrangements have given way to random variables (i.e. the functions that give the payoffs in utility in each possible situation). Additivity of payoffs *in a situation:*

$$U\ [B_1 \# B_2\ (w) = U\ [B_1\ (w)\ ] + U\ [B_2\ (w)\ ]$$

which fails for physical aggregation, holds *by definition* for mathematical aggregation. Then, if we grant (1) The constant betting arrangement, C, which returns utility V in every possible circumstance has expected utility V, $EU(C)$ = V and (2) If $EU\ (B^1) = EU\ (B^{1\prime})$ and $EU\ (B^{2\prime}) = EU\ (\mathrm{B}^{2\prime})$ then $EU\ (B^1 \# B2) = EU\ (B^1 \# B^{2\prime})$ then it follows that expected utility is additive over mathematical aggregation:

$$EU\ (B^1 \# B^2) = EU\ (B^1) + EU\ (B^2)$$

If *p;q* are incompatible the following are descriptions of the same betting arrangement: (I) A bet on *p or q* which returns utility one if *p or q*; utility zero otherwise, (II) The mathematical aggregation of a bet on *p* which returns utility one if *p*; zero otherwise and a bet on *q* which returns utility one if *q*; zero otherwise. Then by additivity over mathematical aggregation; the definition of probability in terms of expected utility; and the principle that a betting arrangement gets the same expected utility no matter how described we get the additivity law for probability.

Allow me to make a brief remark about the Ramsey extensionality principle. Belief is notoriously non-extensional. Why should degree of belief obey an extensionality principle? The answer is, of course, the fact that the two descriptions designate the same betting arrangement is an immediate consequence of the underlying truth functional logic. Ramsey's view is that coherence is a kind of *consistency* and that the theory of personal probability is a branch of *logic*.

Persistent critics of the argument will want to attack (2) above, though it is hard to see how. Since we have passed to utility theory, and mathematical aggregation, there is no way in which payoffs can interact at worlds other than by addition. Since our abstracted "bets" are not events or propositions but random variables there is no way for a "bet" to influence the probabilities salient to the evaluation of an aggregate. For those who are nevertheless still dubious about (2) it is possible to found additivity on more restricted substitution principles within utility theory [e.g. see Anscombe and Aumann (1963) and Raiffa (1968).] Those who are dubious about even the restricted substitution principles should consider what we can do in special cases with (1) alone.

Consider this twist on Raiffa. Suppose that there is a known fair coin, so that CHANCE (H) = CHANCE (T) = ½ is known. Suppose p is an "ethically neutral" proposition in the sense of Ramsey, i.e. that the truth or falsity of p makes no difference to the decision maker no matter what the payoffs. Suppose also that p is such that the decision maker is indifferent between the gambles (i) utility 1 if p; utility 0 if not p, and (ii) utility 1 if not p; utility zero if p. Consider a gamble with the payoff matrix:

	p	not p
H	1	0
T	0	1

This gamble can be represented two ways:

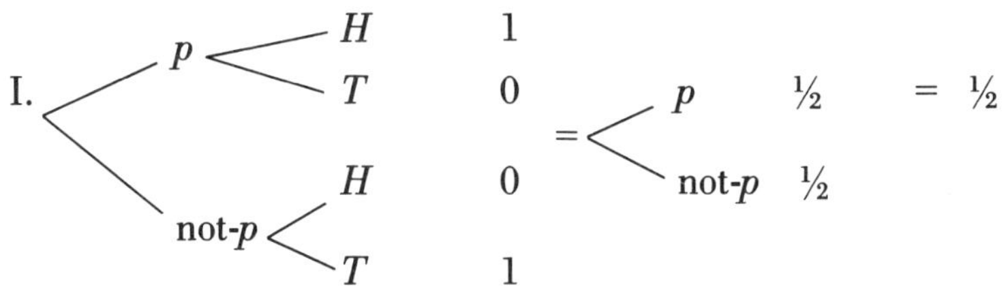

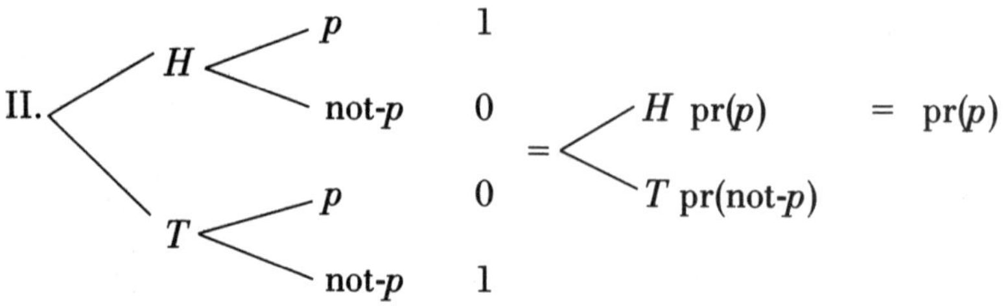

The first equality in I comes from the known chance and definition of utility; the second fom principle (1) page 9. The first equality in II comes from the definition of personal probability of p and the subjects stated indifference; the second from principle (1) page 9. I and II are descriptions of the same betting arrangement. The different orders in I and II are artifacts of the description rather than indicative of any temporal or causal order. So consistency requires that $\text{pr}(p) = \text{pr}(\text{not-}p) = \frac{1}{2}$. Along the same lines we can validate an n-fold equiprobable partition for any n. If a critic lets us get this far, getting the rest of the way is not difficult.

Once we have seen the foregoing, do we really need the physical probabilities as input? If p is ethically neutral and the subject is indifferent between gambles (i) and (ii) we can define $\text{pr}(p) = \text{pr}(\text{not-}p) = \frac{1}{2}$; and similarly for the appropriate generalization to n-fold partitions. If we pursue this line we, like Ramsey, will have followed the dutch book theorems to deeper and deeper levels until it leads to the representation theorem.

This is not to say that the qualitative assumptions of the representation theorems for probability and utility are above question; but rather to emphasize that criticism of the dutch book arguments, if it is to be more than superficial, must question utility theory and the representation theorems as well.

I have spent this much time chewing over the classical dutch book theorems in hopes of generating some sympathy for Ramsey's view that they touch something fundamental. It is time now to survey some newer results regarding dynamic coherence.

III. New horizons

As Kyburg (1978) and Hacking (1967) point out, a theory of personal probability which addresses only coherence at a time, and neglects probability change is seriously incomplete. Kyburg puts it nicely:

> It might be maintained, and would be by anyone who regarded the theory of subjective probability as providing insights into scientific inference, that its main function is dynamic: it is the changes in the probability function that are wrought by empirical evidence, through the mediation of Bayes' theorem (or a generalization thereof) that give the theory its philosophical importance. [Kyburg (1978) 176]

The dynamic rule of probability change that Kyburg has in mind is *Bayes' rule* or the principle of *conditionalization.* That is if one gets as new evidence just the proposition, e, where e is a proposition (measurable set) in ones probability space with positive prior probability, one should update by taking as a new probability for any proposition, q, its old probability conditional on p:

$$Pr^{new}(q) = Pr^{old}(e \,\&\, q)\,/Pr^{old}(p)$$

Both Ramsey and deFinetti endorse the principle of conditionalization:

> . . . obviously if p is the fact observed, my degree of belief in q after the observation should be equal to my degree of belief in q given p before, or by the multiplication law to the quotient of my degree of belief in $p\&q$ by my degree of belief in p. [Ramsey (1931) 192.]
>
> The aquisition of a further piece of information, H—in other words *experience,* since experience is nothing more than the aquisition of further information—acts always and only in the way we have just described . . . As a result of this the probabilities are $P(E/H)$ instead of $P(E)$. [deFinetti (1974) 141] [1]

But nowhere in Ramsey and deFinetti do we find an explicit coherence argument for conditionalization. Kyburg sees this as a deficiency in the theory of personal probability:

> But the really serious problem is that there is nothing in the theory that says a person should *change* his degrees of belief in response to evidence in accordance with Bayes' theorem. On the contrary, the whole thrust of the subjectivistic theory is to claim that the history of an individual's beliefs is irrelevant to their rationality: all that counts at a given time is that they conform to the requirements of coherence. [Kyburg (1978) 176-177]

Hacking (1967) calls it the "dynamic assumption of personalism" and argues that no coherence argument for conditionalization is possible.

deFinetti (1937) does show that there is a coherence argument for the ratio definition of conditional probability:

$$\Pr(q/p) = \Pr(q \& p) / \Pr(p)$$

based on conditional bets, i.e. bets called off if the condition is not fulfilled. Let the conditional probability of q given p be defined as the fair betting quotient for a conditional bet on q on the condition p. A conditional bet on q conditional on p can be represented as the aggregate of unconditional bets on p & q and against p. For the two ways of evaluating the conditional bet to agree, the ratio measure of conditional probability is required. But as Hacking (1967)[2] points out, this is a *static* coherence argument regarding coherence of conditional and unconditional probabilities at the same time; and falls short, by itself, of justifying the rule of conditionalization.

It does not fall far short, however. If you have an epistemic *rule* for updating degrees of belief, and that rule disagrees with the rule of conditionalization, then a dynamic dutch book can be made against you. Omitting details, the heart of the matter is that if a cunning bettor knows your rule for updating degrees of belief, there are two ways in which he can make a bet on q conditional on p with you. He can make a conditional bet now at your going conditional betting ratio, or he can reserve an amount to bet on q if the condition p is satisfied. For the two conditional betting ratios to coincide, your rule must be to update by conditionalization. This argument, due to David Lewis, is reported in Teller (1973) (1976). An elementary exposition is available in the second and third editions of my *Choice and Chance*. I want to emphasize that here again possibility of a dutch book is a symptom of a deeper pragmatic inconsistency: the comittment to two different betting ratios for the same conditional bet.

Lewis' argument for conditionalization fills the gap that Kyburg and Hacking found in the theory of personal probability. Is the theory then complete? One can argue that it is only if the conditions suitable for conditionalization exist whenever we learn from experience, i.e. the results of an observation are always summed up as rendering certain some proposition in the observers probability space. Thus put, the view seems at best extremely dubious. It appears that Ramsey did not hold it. His discussion of conditionalization begins:

> Since observation changes (in degree at least) my opinion about the fact observed, some of my degrees of belief after the observation are necessarily inconsistent with those I had before. We have therefore to explain how exactly the observation should modify my degrees of belief . . .

There follows the endorsement of the rule of conditionalization already quoted. The tantalizing parenthetical phrase, suggestive of a theory of uncertain observation, is not followed up.[3]

Such a theory, where observation raises the probability of a proposition, but does not raise it all the way to one, is developed in Jeffrey (1965) (1968). Jeffrey proposes a generalization of conditionalization on p, where the probabilities conditional on p are kept constant but the probability of p need not go to one. Such a change, he calls a change by probability kinematics on p. More generally, if $\{p_1\}$ is a finite partition, all of whose members have positive prior probability, a change is by probability kinematics on the partition if:

$$\text{Pr}_{\text{new}}\ (q \text{ given } P_i) = \text{Pr}_{\text{old}}\ (q \text{ given } P_i)$$

Jeffrey sees probability kinematics as a more general method of updating suitable for uncertain observations whose informational content is captured by the probability shifts of the members of the partition. He uses the example of the observation of the color of a cloth by candlelight with probability change by probability kinematics on a partition of colors. For a discussion of probability kinematics and its relation to Bayesian statistics, see Diaconis and Zabell (1982).

Is there a coherence argument for probability kinematics? One approach is to try to reduce probability kinematics to conditionalization on a larger space. Thus, if we introduce higher order probabilities, we might suppose that the uncertain observer did learn something for certain—not *p* but rather that the probability of *p* is .99. If we put restrictions on the higher order probability measure that guarantee that pr(*p*) = .99 only carries first order information about *p*, then we get belief change by probability kinematics on *p*. Likewise for the more general form of probability kinematics. [Armendt (1980), Skyrms (1980 a, b), Good (1981)]. For the special case where the partition is {*p*, not *p*} the condition:

SUFFICIENCY: PR[*q* given *p* & pr(*p*) = *a*] = PR[*q* given *p*]

(for first order *p*,*q*) guarantees that conditionalization on pr(*p*) = *a* effects a change on the probability over the first order statements by probability kinematics on the partition {*p*, not *p*}. The additional condition:

MILLER: PR[*p* given pr(*p*) = *a*] = *a*

guarantees that conditioning on pr(*p*) = *a* gets you to a posterior distribution in which the probability of *p* does equal *a*. Roughly speaking, MILLER together with the theorem on total probability gives you:

EXPECTATION: PR[*p*] = EXPECTATION [pr(*p*)]

For more detail see the references. The only difficulty with this approach is that it solves a slightly different problem than the one posed by Jeffrey. He does not suppose that we are presented with

"The Given in probabilistic terms" but rather that there is no proposition which we learn for certain [Jeffrey (1983) 183].

It is, however, also possible to solve the problem in Jeffrey's own terms and give a dutch book argument for probability kinematics using only first order probabilities [Skyrms (forthcoming)]. The argument is again to be made precise as an argument addressed to *rules* or *strategies* for changing degrees of belief, in a certain sort of epistemic situation. The trick is to consider not only the uncertain observation, but also to introduce a subsequent observation which removes the uncertainty.

For definiteness, suppose that the question at issue concerns the color and flavor of a jellybean. The bookie has an initial opinion and posts odds; observes the bean under dim light, and posts odds again; then the lights are turned up (of more definitely, the true color announced) and the bookie posts odds again. The bookie has a *strategy* for dealing with this situation. The strategy can no longer be thought of as a *function* since there is no "given" at the time of uncertain observation, but must rather be thought of as a relation, i.e. as a set of quadruples $<Pr_1, Pr_2, COLOR, Pr_3>$ which delimit his possible moves. There must be, however a way of guaranteeing that under dim light he only observed the color of the bean rather than tasteing it, or more precisely that his *strategy* treats the observation qualitatively as one confined to color. We therefore require that his probability at time 3 after the true color is announced not depend on the results of the imperfect information reflected in Pr_2 but rather be determined by his initial probability together with the color announced. This can be thought of as the requirement that the partition of colors be *qualitatively sufficient with respect to his strategy.* Then we can show that for a suitable notion of coherence, coherence requires a strategy which proceeds from the initial probability to the one modified by uncertain observation by probability kinematics on the partition of colors.

If the bookie has higher order initial probabilities about the possible pr_2s on the color-flavor space that may eventuate from uncertain observation, then we can say more. If the bookie's strategy does not proceed almost everywhere in his pr_1 (i.e. if at the initial time he gives any positive probability at all to the proposition that his moves will violate probability kinematics) the he is subject to an unconditional dutch book. [Skyrms (forthcoming)].

Higher order probabilities also bring in an additional coherence requirement. Goldstein (1983) and van Fraassen (1984) show that coherence requires the aforementioned *Expectation Principle,* that pr_1 be equal to the prior expectation of pr_2. That is, the bookies initial probability over the color-flavor space must be the "center of mass" of the possible probabilities at time two, with the higher order initial probabilities being taken as the measure of mass, or else he lays himself open to an unconditional dutch book.

One can also pursue converse dynamic dutch book theorems in the domains under consideration. Let a call a bookie's strategy in the uncertain observation game with the jelly bean *catholic* if it allows for pr_2 which distribute probabilities over the colors in every posible way consistent with giving each color non-zero probability. Then in that game, a *catholic* strategy of belief change by *probability kinematics* is coherent. [Skyrms (1987)]

IV. Invitation

The first half of this paper was concerned with arguing that dutch book arguments, properly interpreted, touch deep features in the logic of practical reason. The second half—an incomplete survey of recent work in dynamic and higher order coherence—is intended to show that the field is not closed. There is no reason to think that the sort of model situation used to analyze probability kinematics for uncertain observations is the only model of epistemological interest. For example, there is the statistical model investigated by Cornfield (1969), Freedman and Purves (1969), Heath and Sudderth (1972) (1978), Lane and Sudderth (1983) wherein the chance of an experimental outcome given the state of nature is known but no prior probability over the states of nature is assumed, the experiment is performed and the bookie posts odds according to some rule. You can think of other models of interest. The contribution of investigations of coherence to epistemology is an unfinished story.[4]

NOTES

1. And deFinetti (1937): " . . . observation can only give us information capable of influencing our opinion. The meaning of this statement is very precise: it means that to the probability of the fact conditioned on this information—a probability very distinct from that of the same fact not conditioned on anything else—we can indeed attribute a different value." Compare also deFinetti (1976) section VII.

2. I take it that Levi (1980), pp. 81-82 is making the same point in his discussion of "confirmational conditionalization" vs. "temporal credal conditionalization." So is Seidenfeld (1985) 280 ftnt 7. Both Levi and Seidenfeld, like Hacking, are skeptical about the possibility of a dutch book argument for conditionalization.

3. Did Ramsey think about such matters at all? As Diaconis and Zabell (1982) point out, there is a partial anticipation of the "probability kinematics" of Jeffrey (1965) (1968) in Donkin (1851). It is therefore of some interest that Ramsey took notes on Donkin's paper, underlining some of the passages relevant to our concerns here. Richard Nollan, curator of special collections at the Hillman library of the University of Pittsburgh, was kind enough to allow me to examine the Ramsey papers in the Archives for Scientific Philosophy in the Twentieth Century of which the notes in question are item 003-13-01.

4. Research partially supported by National Science Foundation Grant SES-84-09898.

REFERENCES

E. Adams, "On Rational Betting Systems," *Archive fur Mathe-*matische Logik und Grundlagenforshung, vol. 6 (1962), pp. 7-29, 112-128.

E. Adams, and R. Rosenkrantz, "Applying the Jeffrey Decision Model to Rational Betting and Information Acquisition," *Theory and Decision*, vol. 12 (1980) pp. 1-20.

B. Armendt, "Is there a Dutch Book Argument for Probability Kinematics?" *Philosophy of Science*, vol. 47 (1980), pp. 583-588.

F. J. Anscombe and R. J. Aumann, "A Definition of Subjective Probability," *Annals of Mathematical Statistics*, vol. 34 (1963), pp. 199-205.

P. Baillie, "Confirmation and the Dutch Book Argument," *British Journal for Philosophy of Science*, vol. 24 (1973), pp. 393-397.

R. J. Buehler, "Coherent Preferences," *The Annals of Statistics*, vol. 4 (1976), pp. 1051-1064.

C. Chihara and R. Kennedy, "The Dutch Book Argument: Its Logical Flaws, Its Subjective Sources," *Philosophical Studies*, vol. 36 (1979), pp. 19-33.

J. Cornfield, "The Bayesian Outlook and Its Application," *Biometrics* vol. 25 (1969), pp. 617-658.

B. de Finetti, "La Prevision: ses lois logiques, ses sources subjectives," *Annales de l'Institut Henri Poincare*, vol. 7 (1937), pp. 1-68. trans. as "Foresight: its logical laws, its subjective sources" in H. E. Kyburg, Jr. and H. Smokler (ed.), *Studies in Subjective Probability* (Huntington, N.Y.: Krieger, 1980).

B. de Finetti, *Theory of Probability*, vol. I (New York: Wiley, 1974).

B. de Finetti, "Probability: Beware of Falsifications," *Scientia*, vol. 111 (1976), pp. 283-303, reprinted in Kyburg and Smokler (1980).

P. Diaconis and S. Zabell, "Updating Subjective Probability," *Journal of the American Statistical Association*, vol. 77 (1982), pp. 822-830.

W. F. Donkin, "On Certain Questions Relating to the Theory of Probabilities," *The Philosophical Magazine*, Ser. 4, 1 (1851), pp. 353-361.

Z. Domotor, "Probability Kinematics and the Representation of Belief Change," *Philosophy of Science*, vol. 47 (1980), pp. 384-404.

Ellery Eells, *Rational Decision and Causality* (Cambridge: Cambridge University Press, 1982).

D. A. Freedman and R. A. Purves, "Bayes Method for Bookies," *The Annals of Mathematical Statistics*, vol. 40 (1969), pp. 1177-1186.

M. Goldstein, "Revising Previsions: A Geometric Interpretation," *Journal of the Royal Statistical Society B*, vol. 43 (1981), pp. 105-130.

M. Goldstein, "The Prevision of a Prevision," *Journal of the American Statistical Association*, vol. 78 (1983), pp. 817-819.

I. J. Good, "The Weight of Evidence provided by Uncertain Testimony or from an Uncertain Event," *Journal of Statistical Computation and Simulation*, vol. 13 (1981), pp. 56-60.

C. Glymour, *Theory and Evidence* (Princeton: Princeton University Press, (1980), pp. 71-72.

I. Hacking, "Slightly More Realistic Personal Probability," *Philosophy of Science*, vol. 34 (1967), pp. 311-325.

D. Heath and W. Sudderth, "On a Theorem of de Finetti, Oddsmaking and Game Theory," *Annals of Mathematical Statistics*, vol. 43 (1972), pp. 2072-2077.

D. Heath and W. Sudderth, "On Finitely Additive Priors, Coherence, and Extended Admissibility," *The Annals of Statistics*, vol. 6 (1978), pp. 333-345.

F. Jackson and R. Pargetter, "A Modified Dutch Book Argument" *Philosophical Studies*, vol. 29 (1976), pp. 403-407.

R. Jeffrey *The Logic of Decision* (N. Y.: McGraw Hill, 1965); 2nd ed. (Chicago: University of Chicago Press; 1983).

R. Jeffrey, "Probable Knowledge," in I. Lakatos (ed.), *The Problem of Inductive Logic* (Amsterdam, North Holland, 1968), reprinted in Kyburg and Smokler (1980).

M. Kaplan, "Decision Theory as Philosophy," *Philosophy of Science* vol. 50 (1983), pp. 549-577.

J. Kemeny, "Fair Bets and Inductive Probabilities," *Journal of Symbolic Logic*, vol. 20 (1955), pp. 251-262.

H. Kyburg, "Subjective Probability: Criticisms, Reflections and Problems," *The Journal of Philosophical Logic*, vol. 7 (1978), pp. 157-180.

H. Kyburg and H. Smokler (eds.), *Studies in Subjective Probability* (Huntington, New York: Krieger, 1980).

D. Lane and W. Sudderth, "Coherent and Continuous Inference," *The Annals of Statistics*, vol. 11 (1983), pp. 114-120.

R. Lehman, "On Confirmation and Rational Betting," *Journal of Symbolic Logic*, vol. 20 (1955), pp. 251-262.

Isaac Levi, *The Enterprise of Knowledge* (Cambridge, Mass.: MIT, 1980).

H. Putnam, "Probability Theory and Confirmation," in *Mathematics, Matter and Method* (Cambridge: Cambridge University Press, 1975).

H. Raiffa, *Decision Analysis* (Reading, Massachusetts: Addison-Wesley, 1968).

F. P. Ramsey, "Truth and Probability," in R.B. Braithwaite (ed.), *The Foundations of Mathematics and Other Essays* (N.Y.: Harcourt Brace, (1931) and in Kyburg and Smokler (1980).

F. Schick, "Dutch Books and Money Pumps," *The Journal of Philosophy*, forthcoming.

A. Shimony, "Coherence and the Axioms of Confirmation," *Journal of Symbolic Logic*, vol. 20 (1955), pp. 1-28.

A. Shimony, "Amplifying Personal Probability," *Philosophy of Science*, vol. 34 (1967), pp. 326-333.

T. Seidenfeld, "Calibration, Coherence and Scoring Rules," *Philosophy of Science*, vol. 52 (1985), pp. 274-294.

T. Seidenfeld and M.J. Schervish, "A Conflict between Finite Additivity and Avoiding Dutch Book," *Philosophy of Science*, vol. 50 (1983), pp. 398-412.

B. Skyrms, *Causal Necessity* (New Haven: Yale University Press, 1980a) Appendix 2.

B. Skyrms, "Higher Order Degrees of Belief,"in D.H. Mellor (ed.), *Prospects for pragmatism* (Cambridge: Cambridge University Press, 1980b).

B. Skyrms, "Zeno's Paradox of Measure" in R.S. Cohen and L.

Lauden (ed.), *Physics, Philosophy, and Psychoanalysis (Dordrecht: Reidel, 1983), pp. 223-54.*

B. Skyrms, Pragmatics and Empiricism (New Haven: Yale University Press, 1984).

B. Skyrms, *Choice and Chance: An Introduction to Inductive Logic* (Belmont, California: Wadsworth, 1986).

B. Skyrms, "Dynamic Coherence and Probability Kinematics," *Philosophy of Science*, forthcoming, 1987.

S. Spielman, "Physical Probability and Bayesian Statistics," *Synthese*, vol. 36 (1977), pp. 235-269.

P. Teller, "Conditionalization and Observation," *Synthese*, vol. 26 (1973), pp. 218-258.

P. Teller, "Conditionalization, Observation, and Change of Preference," in W. Harper and C. Hooker (eds.) *Foundations of Probability Theory, Statistical Inference, and Statistical Theories of Science* (Dordrecht-Holland: D. Reidel, 1976), pp. 205-253.

B. van Fraassen, "Belief and the Will," *Journal of Philosophy*, vol. 81 (1984), pp. 235-256.

B. van Fraassen, "Rational Belief and Probability Kinematics," *Philosophy of Science*, vol. 47 (1980), pp. 165-187.

J. Vickers, "Some Remarks on Coherence and Subjective Probability," *Philosophy of Science*, vol. 32 (1965), pp. 32-38.

J. von Neumann and O. Morgenstern, *Theory of Games and Economic Behavior* (Princeton, New Jersey: Princeton University Press, 1944).

P.M. Williams, "Bayesian Conditionalization and the Principle of Minimum Information," *British Journal for the Philosophy of Science*, vol. 31 (1980), pp. 131-144.

PEIRCE MEETS KUHN: THE VERIFICATIONIST THEORY OF TRUTH WITHOUT A FIXED METHODOLOGY OF VERIFICATION

ARON EDIDIN

VERIFICATIONISM, that hardy perennial that seems to outlive each of the philosophical doctrines thought to provide its broad theoretical justification, is enjoying yet another renaissance, thanks in large part to the work of Dummet and Putnam.[1] Dummett seems chiefly concerned to defend an account of meaning as determined by verification-conditions, with a verificationist theory of truth a significant corollary. For Putnam, on the other hand, verificationism about truth plays the central role. In what follows, I shall be concerned exclusively with the consequences of adopting a verificiationist account of truth.

I. The verificationist account of truth

On the sort of account I have in mind, truth, as Putnam puts it, is "an idealization of rational acceptability."[2] Roughly, a statement is true if and only if it would ultimately be accepted by an ideal community of rational investigators. The classic statement of a view of this kind is Peirce's:

> The opinion which is fated to be ultimately agreed to by all who investigate, is what we mean by the truth, and the object represented in this opinion is the real . . .

> Our perversity and that of others may indefinitely postpone the settlement of opinion; it might been conceivably cause an arbitrary poroposition to be universally accepted as long as the human race should last. Yet even that would not change the nature of the belief, which alone could be the result of investigation carried sufficiently far . . . the reality of that which is real does depend on the feal fact that investigation is destined to lead, at least, if continued long enough, to a belief in it.[3]

To be sure, my rough formulation of the verificationist account of truth can't be quite right. One problem, as Plantinga has noted, is that if, as the account entails, an ideal community of ideally rational inquirers in the long run accepts all and only truths, such a community must ultimately accept the statement that an ideal community of ideally rational inquirers exists. The existence of such a community (henceforth called an "irc") would thus follow from the very definition of truth.[4] This is an unwelcome consequence, at least for those of us who think that no irc exists and that none is likely to arise. The need to modify the rough definition to avoid this consequence poses a serious challenge to defenders of a verificationist approach to truth. But I think that the rough definitions will suffice for my purposes.

Now, Peirce thought that every meaningful question had an answer that an irc would ultimately accept.[5] He could thus combine a verificationist account of truth with belief in the principle of bivalence, i.e., the claim that every meaningful statement is either true or false, where a statement is false just in case its negation is true. But if we don't share Peirce's optimism concerning the capcity of rational inquiry to answer every meaningful question, this combination of views will not be available to us. A verificationist view of truth will commit us to the view that some statements are neither true nor false. In terms of formal propositional logic, this means that the Peircean verificationist can adopt the standard classical truth tables for the propositional connectives, while more pessimistic verificationists must admit possibilities not covered by those tables. Whether rejection of bivalence and the classical truth tables requires a weakening of the classical relation of logical consequence is a question to which I shall return.

II. Ideal rationality

The verificationist account of truth as I have described it appeals to the notion of an ideal community of ideally rational inquirers. How ought we to understand ideal rationality in this context? Without attempting a detailed answer, we may at least reject some candidates which would not too long ago have been very popular. As many have recognized, one profound lesson of Kuhn's work on the nature of theory choice in science is that the rationality embodied, however imperfectly, in scientific reasoning is not a matter of possessing and applying correct algorithms for deriving conclusions from the available evidence. Ideal rationality ought *not* to be understood as a matter of following some fixed set of methodological rules that generate a unique rational set of beliefs for any possible body of evidence. Instead, our idea of rationality must allow for such factors as individual judgement, preferences among such intellectual values as simplicity, comprehensiveness, and evidential support, and respect in varying degrees for the judgement and intellectual values of other inquirers.

The point may be made more vivid by considering what would happen if Tinkerbell were to wave her magic wand and turn us all into Ideally Rational Inquirers. How would we be changed? To begin with, I suppose that most of us would become much more intelligent than we now are. We'd be able to understand longer sentences and more complex ideas than we can now grasp; perhaps there would be no limit to the complexity of ideas we could understand. In addition, we'd get to be much better at recognizing the logical consequences of propositions; perhaps we'd even become logically omniscient, aware of all logical truths and entailments in all systems of logic. Moreover, we'd be freed of any number of prejudices. We would no longer be inclined to favor our own hypotheses[6] and those of our friends and compatriots over those of strangers, foreigners, or academic or ideological rivals. If Tinkerbell made our circumstances as well as our persons ideal for rational inquiry, resources for inquiry would be inexhaustible, political and ideological interference nonexistent, and communication easy and instantaneous among all inquirers. Thus a truly ideal community of inquirers would include all inquirers everywhere.

On the other hand, there is no reason to think that all of the personal idiosyncracies that influence the acceptance of hypotheses and theories would disappear or become universal. Surely some of us would remain more inclined than others to go beyond more direct sorts of evidential support to accept hypotheses of broad explanatory power. We would continue to differ in our predilections for or against reductive analyses, in our tastes for ontological economy or plenitude, in viewing common sense as something to be accomodated or transcended.

But if the history of science teaches us that rational inquirers can differ in these ways, it also teaches us that a community of rational inquirers is likely in time to achieve a broad consensus in spite of such differences.[7] Perhaps, as Lehrer suggests, a minimal degree of mutual respect among inquirers should be sufficient to generate consensus.[8] In any case, it seems reasonable to expect that if we were transformed into an ideal community of ideally rational inquirers we would in the long run achieve consensus concerning most of the questions we considered.

III. Interlude: truth by convention

If truth is to be coextensive with what an irc would accept, we must suppose that an irc would investigate *everything*. Were Tinkerbell to transform us into an irc, there would be no limit to the matters about which we would draw conclusions. Among innumerable other things, we would draw conclusions about our own history. For every statement that we ultimately accepted, there would presumably be a story about *why* we accepted it. Most of the time, perhaps, the story would be a simple tale of discovery, such as "On February 11, 1984, those of us who were in South Bend, Indiana, saw that it was cloudly all day." or "Electric charge is quantized. The Millikan experiement showed that the only observed charge is always an integral multiple of nature's basic quantum of charge, the elementary charge *e*."[9] In other cases, while not a simple discovery, the adoption of a statement might be explained as the result of an accumulation of evidence in its favor, or of complex reasoning based on available evidence.

Sometimes, though, the story might refer not to discovery or evidence but to a conventional decision in which pragmatic considerations play a prominent role. Here, for example, is a story about how we came to accept a specification of the wavelength in meters of krypton-86 light:

> The meter was first chosen so that the distance from the earth's pole to the equator along the meridian line through Paris would be 10,000,000 m. Later the meter was defined as the distance between two fine scratches on a carefully preserved bar of platinum-iridium. At present the meter is defined in terms of the wavelength of the orange-red radiation in vacuum corresponding to the unperturbed transition between the levels of $2p_{10}$ and $5d_5$ of the atom krypton-86:
>
> 1m = 1,650,763.73 wavelengths of krypton-86 light.
>
> A standard of length must meet the requirements of easy accessibility and high accuracy, unappreciably influenced by such disturbances as changes in temperature or pressure. The platinum-iridium bar, which was the standard of length before 1960, can expand with a temperative rise or shrink with a pressure rise. This leads to ambiguities compounded by the width of the scratches. And, worst of all, the bar can be lost or destroyed!
>
> The krypton-86 spectral line overcomes these difficulties in large measure. The wavelength of this electromagnetic radiation is (nearly) independent of external changes and depends only on the structure of the atoms of krypton-86, which is, of course, identical for all such atoms.[10]

One can imagine similar stories being told about the adoption of classical logic in preference to weaker alternatives, the identification of the ordinal numbers with von Neumann's set-theoretic ordinals rather than with Zermelo's or neither, the choice of Riemannian geometry in preference to Euclidean geometry plus universal forces, and so on. I don't mean to defend the claim that an irc would in fact make the choices I've mentioned, much less that it would in fact explain those choices as conventional decisions rather than as discoveries. Much ink has been spilled over the availability of evidence which could decide such issues in logic and mathematics. What I do want to claim is that the possibility of such an explanation is quite intelligible. If I'm right, then the verificationist theory of truth yields a very clear and simple notion of truth by convention: a statement is true by convention just in case it would appear in the Ultimate Theory of an irc and its adoption and retention by the irc would be explained in that theory as the result of conventional decision rather than recognition of evidential support.

Note that this clear and simple definition of truth by convention doesn't entail that anything *is* true by convention. Maybe no irc would continue indefinitely to accept any statement in the absence of convincing evidence in its favor. Maybe the account of evidence that an irc ultimately would adopt would count intuitively pragmatic considerations as genuinely evidential. (That is to say, maybe it wouldn't be rational to continue indefinitely to accept a proposition in the absence of convincing evidence in its favor, and maybe intuitively pragmatic considerations really are genuinely evidential.[11]) But it doesn't seem far-fetched to see traditional arguments for the truth by convention of one or another statement as claiming *at least* that it is rational to accept and continue indefinitely to accept the statement in question even though there is no convincing *evidence* in its favor.

IV. IRC SINGULAR OR IRCS PLURAL?

So far, I've been taking at face value the verificationist claim that (roughly) a statement is true if and only if it would ultimately be accepted by an ideal community of ideally rational inquirers. But an ideal community of ideally rational inquirers could take any of various forms, as could its members. Clearly, ircs could differ in any respects not relevant to their inquiries: some could consists of carnivores, others of herbivores, some could contain more music-lovers than sports fans while others contained more sports fans than music-lovers, and so on. If these were the only possible sorts of variation among ircs, reference to "*an* irc" in the standard formulation of the verificationist account of truth would pose no special problems. If the only variation among ircs would be irrelevant to their inquiries, then if one irc would ultimately accept some statement, so would they all. But if other sorts of variation are possible, this convenient unanimity is thrown into doubt.

If my earlier discussion of ideal rationality is on the right track, then ircs *can* differ in ways that are very relevant to their inquiries. I suggested that ideal rationality for individuals is compatible with any of a wide variety of what might be called "intellectual styles" or sets of preferences among such epistemic values as simplicity, direct evidential support, ontological parsimony or plenitude, and respect for common sense and the opinions of colleagues and peers. Similarly, it seems that ideal rationality in a community is compatible with any of various distributions of intellectual styles among its members. There doesn't seem to be a single rationally ideal mix of conservatives and iconoclasts, reductionists and anti-reductionists, simplifiers and complicators, the timid and the bold in accepting hypotheses beyond those with the more direct sorts of evidential support. Now, the mix of intellectual styles within an irc will surely influence the course of that irc's inquiries,[12] so ircs will differ with respect to what statements they would accept at various stages of their inquiries. It doesn't automatically follow that ircs could differ with respect to what statements they would *ultimately* accept, but the multiplicity and variety of possible ircs makes even this stronger conclusion quite plausible.

Indeed, there are reasons for holding that stronger conclusion which don't depend principally on the various mixes of intellectual styles among possible ircs. If there are questions which an irc would answer not by appeal to convincing evidence but rather on the basis of broadly pragmatic considerations, then it may be what answer is ultimately chosen could depend in part on the *non*-epistemic values of the members of the community.

When variability among ircs with respect to their members' epistemic values is recognized, the case for the possibility of variation with respect to their ultimate conclusions is strengthened. Clearly, the intellectual values predominating in a community will help to shape that community's choices where those choices aren't constrained by the availability of convincing evidence. Even where a conclusion is reached on the basis of the available evidence, it seems plausible that different ircs might differ in their ultimate appraisal of the evidence. Quine has suggested that all the possible evidence past, present, and future could yet leave two (or more) incompatible hypotheses tied for the status of best confirmed. Once we recognize that confirmation involves the invocation of a number of epistemic values whose relative weight may be a matter of legitimate difference both among individuals and among ircs, it seems that the judgement of whether there *is* ultimately a tie between hypotheses may differ from irc to irc, and among those which hold that their is no tie, some may prefer one alternative and others another.

If we take seriously the possibility that ircs could differ with respect to their ultimate conclusions, the original formulation of the verificationist definition of truth must be modified. We ought to say, roughly, that a statement is true if and only if it would ultimately be accepted by *any* ideal community of ideally rational inquirers.

On this account, a statement that some ircs would accept but that others would not accept is not true. An alternative, apparently favored by Goodman,[13] would be to say that a statement is true if and only if it would be accepted by *some* irc. This would require either admitting logically incompatible elements into the ultimate True Account of Everything or else—the option Goodman adopts—relativising the notion of truth from the outset. I find neither option attractive, though I do think that the notion of absolute truth as what any irc would accept leaves room for a plethora of natural relative notions of truth, to which I shall turn shortly.

V. Truth values and logic

It is possible to hold a verificationist account of truth and yet to believe that every statement is either (absolutely) true or (absolutely) false, if one is confident that ideal inquiry must yield a single determinate answer to every question that can be framed. The same combination is available if one holds a verifiability criterion of meaningfulness according to which a putative statement which would neither be accepted by every irc nor rejected by every irc is meaningless, and therefore no real statement. (In either case, one would be led to the conclusion that ircs can't differ in their ultimate conclusions. This is clear in the first case; in the second it follows from the fact that an irc presumably would never ultimately accept a meaningless pseudo-statement!) Peirce, who did accept both a verificationist account of truth and the principle of bivalence, apparently based his belief in their compatibility *both* on optimism concerning the capacity of science to answer questions *and* on a sort of verifiability criterion of meaningfulness.

On the other hand, neither Peircean optimism about the results of inquiry nor verificationist theories of meaning seem very plausible. It seems likely, then, that if the verificationist account of truth is correct, then some statements are neither true nor false. I've already suggested that ircs might differ in what they ultimately accept. If we say that a statement is false just in case all ircs would ultimately reject it, then a statement that some ircs would ultimately accept and others ultimately reject is neither true nor false on the verificationist account. Moreover, there's reason to suspect that there are some statements that some or all ircs would neither accept nor reject. A promising candidate for such status might be the Coninuum Hypothesis in set theory; more controversial candidates might be statements that attribute determinate positions and momenta to electrons.

So if we adopt a verificationist view of truth, we ought to allow that some statements may be neither true nor false. Moreover, among those statements which are neither true nor false, there is a natural ranking which places some of the closer to being true than others. Suppose we say, following Chisholm,[14] that an irc *withholds* any statement that it neither accepts nor rejects. A true statement is one that every irc would ultimately accept. It seems plausible that there be some statements that some ircs would accept and others withhold, but none reject. Such statements will be closer to being true than, for example, statements that every irc would withhold, and the negations of the former, which some ircs would reject and the rest withhold, would be closer to falsehood. On a par with statements that every irc would withhold are those that some would accept and the rest reject, and those which some would accept, some reject, and some withhold. Schematically:

	Accept	*Withhold*	*Reject*
True	All	None	None
2nd Best	Some	Some	None
Middle	None	All	None
	Some	None	Some
	Some	Some	Some
2nd Worst	None	Some	Some
False	None	None	All[15]

If the classical truth values of truth and falsehood are characterized respectively as acceptance by all ircs with rejection or withholding by none and rejection by all ircs with acceptance or withholding by none, it seems reasonable to consider the five other possible distributions of acceptance, rejection, and withholding to be truth values as well, with the truth values grouped into what could be called five truth levels.[16] This suggests that the semantics for propositional logic will have to be given in some way other than by the classical bivalent truth-tables. In fact, there is reason to suspect that the semantics for propositional logic won't be truth-functional at all. For example, let *P* and *Q* be logically and nomologically unrelated statements such that any irc would withhold both. Then if the semantics for propositional logic is truth-functional, it follows that (*P*&-*P*) and (*P*&-*Q*) must have the same truth value. But it might well be that every irc would reject (*P*&-*P*) while at least some would withhold (*P*&-*Q*).

We can characterize validly under these conditions by saying that a rule of inference is valid just in case every argument of that form is such that its conclusion occupies at least as high a truth level as the conjunction of its premises. Presumably every irc would accept any valid rule. If we suppose (as seems plausible) that the set of statements an irc would ultimately accept ought to be closed under the logical inference-rules the irc would ultimately accept, then the set of rules that generate all and only valid arguments will be the set of those rules that any irc would accept. If, whether because of evidence in its favor or because of its practical advantages, every irc would ultimately accept the inference-rules of classical logic, then the valid argument forms will just be the classically valid ones.

It might be thought that since it's the set of universally accepted inference-rules that generates the class of valid inferences, every irc would accept just this set—any stronger one would generate some invalid inferences. So perhaps any logical inference-rule that could be rejected by *some* irc would be rejected by every irc. On the other hand, a rule which isn't truth-preserving with respect to the verificationist notion of absolute truth may yet preserve some kind of relative truth, and this might be enough to render it acceptable to some irc. So it is to notions of relative truth that we now must turn.

VI. Relative truth

A statement is true on the verificationist account if it would ultimately be accepted by any irc. What about a statement that would ultimately be accepted by any irc with certain special characteristics? Such a statement may not be true, but it seems natural to say that it is true *relative to those characteristics*. This gives rise to a multitude of kinds of relative truth. Among them are:

Truth in a version. This is Goodman's relativistic notion of truth.[17] A version may be taken to be an acceptable Ultimate Theory of Everything.[18] A statement will be true in a given version just in case it would ultimately be accepted by any irc that would ultimately accept that version, i.e., just in case it is contained in that version. Every statement that would be accepted by some irc is true in at least some versions. Note that if an irc's Ultimate Theory is closed under the irc's chosen rules of logical inference, those rules will automatically preserve truth in the irc's version.

Truth given certain assumptions. If a certain set of statements would ultimately all be accepted by some irc, a statement will be true relative to that set just in case it would ultimately be accepted by any irc that ultimately accepted all the statements in the set.

Truth relative to a set of conventions. If a certain set of statements would ultimately all be accepted by some irc as a result of conventional decisions, a statement will be true relative to those conventions just in case it would ultimately be accepted by any irc that made those conventional decisions. It may be that most of the statements that are said to be true by convention are true in only this relative sense. (This is entirely compatible with the possibility that some statements would be accepted by conventional decision by every irc, and so are absolutely true by convention.)

True for you. Even this bane of undergraduate philosophy classes can be given a perfectly good sense on the verificationist account of truth. Among the ircs will be some which more than any others reflect your intellectual characteristics (or at least those characteristics which aren't intellectual flaws). A statement will be true for you just in case it would ultimately be accepted by every such irc. Although most students who speak of what's "true for you" seem to be using the phrase merely to designate what you believe, I suspect that some of the more sophisticated of them have something like this verificationist sense at least vaguely in mind.

A verificationist account of truth yields any number of coherent and natural kinds of relative truth. But does it support any interesting sort of relativism? Yes, in that what's true for me may not be true for you, what's true given one set of assumptions may not be true given another, and so on. More radically, a case can be made for the claim that in asserting a statement we commit ourselves not to its truth but only to its relative truth. This suggestion is supported by the fact that an irc can ultimately accept, and its ideally rational members thus ultimately be willing to assert, statements which some other ircs would not accept and which are thus not true. The most likely candidate for what a speaker is normally committed to may be the "truth for *us* (i.e., the speaker and her intended audience)" of her assertions, where "true for us" is understood to mean ultimately accepted by every irc that maximally resembles what we're like as a group.

Concerning the radical relativist claim that *no* statement is absolutely true, the situation is less favorable. The version of the verificationist theory of truth that I've been concerned with presupposes that there is always a matter of fact concerning what all or some ircs would ultimately accept, so that statements of the form "Statement *S* is true" or "Statement *S* is false" are all themselves either absolutely true or absolutely false. The idea is that ideally rational inquirers would ultimately agree about where rational inquirers could disagree and where they could not.

The absolute truth or falsehood of truth-ascriptions combined with the variability among the Ultimate Theories of the various ircs entails that some instances of Tarski's famous *T*-schema are not true. The schema, where *S* is the statement that *p*, is

S is true if and only if *p*.

If *S* is a statement that some ircs would ultimately accept and other would not, then an irc which would ultimately accept *S* would accept the right-hand side of the biconditional and reject the left-hand side, and so would reject the biconditional. If every irc would withhold *S*, then every irc would withhold the right-hand side and reject the left-hand side and thus the biconditional, so that those instances of the *T*-schema would be false. Note, though, that the *T*-schema would retain this grain of truth: if either side of the biconditional is *true*, the other side must be true as well. And a relativized version of the schema will be true in all cases:

S is true in version *V* if and only if, according to *V*, *p*.[19]

This amounts to the truism that *S* is true in *V* if and only if *V* contains *S*.

In conclusion, the verificationist account of truth seems to require that we reject bivalence and some instances of the unrelativized *T*-schema and accept as meaningful the notions of truth by convention and relative truth along with the possibility that what is true for me may not be true for you. Whether this should be taken to reduce the verificationist theory of truth to absurdity or to recommend it on the basis of its fertility I leave to the ultimate decision of the ircs.

VII. Questions and tentative answers

I have described some characteristics and consequences of what might be called a Kuhnian version of the verificationist theory of truth. No remotely exhaustive account of the advantages and disadvantages of such a theory has been attempted, and numerous important questions have been ignored. The following treats briefly and tentatively some questions that have arisen in discussion of the foregoing. My memory, unfortunately, does not permit me to pair the questions with the questioners who raised them.

(1) *My version of the verificationist theory of truth depends on the assumption that ideally rational inquirers would ultimately agree about where rational inquirers could disagree and where they could not. Why think they would agree about that?*

The first thing to note in response to this question is that the operative notion of rationality here is a very weak one. Just as ircs could ultimately differ with respect to the substance of the theories that they would ultimately adopt, they could also differ with respect to the methodological precepts that they would ultimately recommend. As Kuhn has shown, the details of methodology develop along with theory, and subject to the same sociological influences. The differences in intellectual styles that could lead different ircs in different theoretical directions would do so by inducing the observance (and, presumably, overt espousal) of divergent methodologies. Indeed, as their first description shows, the intellectual styles themselves might as aptly be called methodological predispositions.

My version of the verificationist theory of truth requires that all ircs ultimately agree about what is true and what false despite ultimate disagreement among their favored theories. But this agreement does *not* in turn require ultimate agreement about the details of scientific methodology. Ultimate agreement among all ircs is needed only with respect to what's allowed by the most minimal requirements of general rationality. It might plausibly be supposed that ideal rationality entails both perfect conformity to those requirements and the ability to identify with perfect accuracy instances in which they are violated. If this supposition is correct, then irchood *will* entail ultimately correct judgments of where rational disagreement is ultimately possible, and thus ultimate agreement on these matters with all other ircs.

Verificationists characteristically presuppose the ultimate objectivity and, thus, decisive verifiability, of judgements of verification. On the other hand, it would be desirable to explore the possibility of developing a version of verificationism which does not share this presupposition. This might seem especially appropriate for verificationists who follow Kuhn in asserting the interdependence of methodology and substantive theory and so reject the view that the minimal requirements or rationality include adherence to a quite detailed and specific "scientific method." Perhaps even the most minimal requirements of rationality *are* as much up for grabs as is substantive methodology. Clearly this is an area in which more work needs to be done.

(2) *If you allow ircs to vary with respect to the mixes of intellectual styles within them, their Ultimate Theories of Everything could come out anyhow, especially given the minimal notion of rationality in question. Why think that anything at all (or anything at all interesting) is actually true on this version of the verificationist theory of truth?*

On the theory of truth we're considering, it will be an open question whether anything interesting will turn out to be true, though it may be that the theory will guarantee that there are at least some truths (see Question 1 above). This seems to me appropriate for a Kuhnian verificationist theory. On the other hand, I also think that it is appropriate to allow our judgements about what's clearly true to influence our beliefs about what any irc must ultimately accept. For example, it seems reasonable to hold that the members of an irc must be at least as good as we are at recognizing what is in front of their faces. If we think it's obvious that we can see, e.g., that there are rocks, we should think that the members of any irc would see that as well. In general, if we think that there's compelling evidence for some claim, we ought to think that any irc would appreciate that evidence. This is, of course, not to say that we should take our judgements about what's evident to be infallible or incorrigible. (To the extent that it's obvious that in our most rational moments we can accurately judge egregious violations of the minimal requirements of rationality, this lends further plausibility to the first response to Question 1.)

(3) *If various ircs would ultimately adopt very different-looking Ultimate Theories of Everything, they might well wind up with very different sets of basic concepts. In that case, how could we compare their theories to locate agreement and disagreement?*

If the latter part of the answer to Question 2 is on the right track, it's likely that the differences among the ircs' ultimate theories would be set against a background of substantial agreement against which disagreements could be located. But as the first part of that answer suggested, the theory of truth should not actually require that such agreement obtain. We should, then, allow that different ircs might indeed wind up with theories cast in terms of very different sets of basic concepts. Suppose, then, that an irc ends up accepting a theory which includes no mention of attractive or repulsive forces in its physics. In the absence of further specification of the theory it seems natural to say that this irc ultimately rejects the claim there is such a thing as gravity. But other elements of the irc's story about why we don't fly off the earth may suggest that the irc really does accept the existence of gravity, albeit under some other name or description. How could this question be settled? (Note that its settlement may be crucial to determining whether or not it is *true* that there is gravity.)

The correct response, I think, is that the ircs in question would themselves answer the question explicitly. An irc's ultimate theory is an ultimate theory of *everything*. One of the questions to which such a theory will address itself is: What is the correct description of fundamental physical entities and processes? The answer to this question is the irc's ultimate physical theory, which in the case we're considering, makes no mention of gravity or any other such force. But another question to which the theory will address itself is simply: Is there such a thing as gravity? (Even if the irc would not recognize the existence of gravity itself in its ultimate physics, it would perforce recognize the existence of theories of gravitation in its history of science.) So the question of whether or not this irc ultimately accepts the existence of gravity would after all be answered by the explicit content of its ultimate theory. If that theory contains the claim that there is gravity then, of course, the irc ultimately accepts that claim, and contrariwise if the theory contains the claim that there is no such thing. If the theory contains neither of these claims, then the irc ultimately neither accepts nor rejects the existence of gravity. (This, you will recall, is a possible ultimate response to a claim.)

(4) *Every irc would ultimately know all about what each other irc would ultimately accept. Wouldn't it be possible for each to engage in imaginary dialogue with the others in such a way that*

even though there would be no actual interaction between the ircs, such interaction could be accurately simulated by each? If this were so, wouldn't the pressures that facilitate consesus within ircs be replicated between ircs? And if that were so, wouldn't it make sense to speak once again of one irc instead of many?

If the factors that facilitate consensus within ircs were all matters of what might be called the pure force of the evidence, then imaginary interaction might produce consensus as efficiently as real interaction. But other factors are involved. Such institutional features as standardized scientified education and peer review have no analogues in imaginary dialogue. The same is true of pressures toward agreement arising from common work on joint projects and the use of generally accepted results (whether or not of the status of paradigms in the community in question) as starting-points for further inquiry. All in all, it seems to me that a Kuhnian appreciation of the role of communities of inquiry as such in guiding the progress of that inquiry requires that we take such communities to be genuine groups of really interacting individuals, and this forces a Kuhnian version of Peircean verificationaism to recognize a multiplicity of possible ircs.

NOTES

1. See, e.g., Michael Dummet, *Truth and Other Enigmas* (Cambridge; Mass.: Harvard University Press, 1978), and Hilary Putnam, *Reason, Truth and History* (Cambridge: Cambridge University Press, 1981).

2. Putnam *op cit.*, p. 55.

3. C.S. Peirce, "How to Make our Ideas Clear," in *Philosophical Writings of Peirce* (New York: Dover, 1955), p. 38–9. Peirce diverges from my rough formula in that he makes no mention of ideal rationality in his account of truth. On the problematic role of ideal conditions in Peirce's account cf. Robert Almeder, "Peirce's Thirteen Theories of Truth," *Transactions of the Charles S. Perice Society,* vol. 21 (1985), pp. 77–94.

4. Alvin Plantinga, "How to be an Anti-Realist," *Proceedings and Addresses of the American Philosophical Association,* 1982, pp. 64–66. Plantinga credits Ernest Sosa with the idea for the argument he develops.

5. This belief is expressed in Peirce, *op. cit.,* pp. 38–40.

6. I.e., hypotheses that we have thought up ourselves (as opposed to hypotheses that we accept, which of course we favor).

7. The remarkable ubiquity of such convergence is discussed at length in Larry Laudan, *Science and Values* (Berkeley: University of California Press, 1984).

8. See, esp. Keith Lehrer and Carl Wagner, *Rational Consensus in Science and Society* (Dordrecht: Reidel, 1981).

9. Richard T. Weidner and Robert L. Sells, *Elementary Classical Physics*, Vol. II, (Boston: Allyn and Bacon, 1973), p. 478.

10. *Ibid.*, Vol. I, p. 3.

11. This claim is pressed in Putnam, "The Refutation of Conventionalism," *op. cit.*

12. As might differences in the social and profession organizations of the ircs.

13. See esp. Nelson Goodman, *Ways of Worldmaking* (Indianapolis: Hackett, 1978), Ch. VII. In spite of some disagreements which may be more verbal than substantial (but note Goodman's suspicion of the verbal/substantial distinction, expressed in the cited chapter and elsewhere), Goodman's defense of relativism in this chapter has been the principle influence on my own thought concerning both the possibility of ultimate disagreement among ircs and relative truth.

14. As in Roderick Chisholm, *Theory of Knowledge* (Englewood Cliffs, N.J.: Prentice Hall, 1966), p. 18.

15. It might seem natural to take into account the percentage of ircs taking each attitude toward a statement. Apart from the difficulty in getting any handle at all on the question of what an irc would be *most likely* to do, I suspect that because of the unlimited possibilities for variation in irrelevant respects among ircs, an equal (and infinite) number will accept (reject, withhold) any statement accepted (rejected, withheld) by at least some.

16. To Haack's intriguing question of why such neither-true-nor-false statusses should be considered truth-values the verificationist theory of truth allows a straightforward answer: they are in obvious respect statusses of the same kind as truth and falsehood. Cf. Susan Haack, *Philosophy of Logics* (Cambridge: Cambridge University Press, 1978), pp. 213–15.

17. Goodman, *op. cit*. The verificationist definition of absolute truth corresponds to Goodman's proposed "innocuous interpretation of necessity," *ibid.*, p. 120.

18. I.e., one that would be accepted by some irc.

19. A point also made by Goodman *op. cit.*, p. 120

THE LIFE AND WORK OF THE EARLY CARNAP

NEIL TENNANT

I. A PROFILE OF CARNAP'S PUBLICATIONS

WHAT strikes one most forcibly on working one's way into the chronological profile of Carnap's career was how late he started.

The second thing that strikes one is how much of his extensive early output remains in the original German (and French).

The third thing that struck me was how his interests shifted when he left Europe to come to America, and began publishing only in English. The work in semantics and on probability and induction stands at some remove from his earlier work on other (albeit related) themes. Despite the fact that his most important early works have now been translated, it is likely that many of us know of Carnap's writings only from his American phase. Even if we have studied the earlier works in translation, there is much to be filled in from the untranslated work. A glance at the bibliography in the Schilpp volume reveals just how extensive it is.

For the purposes of retrospective analysis, I would divide his work during his earlier, European period, into the following areas and phases:

The early pre-occupation with philosophical foundations of geometry and physics; in parallel, and through the 'twenties, concern with the problem of the categories, from his own peculiar constructionalist standpoint. Associated with that is the hard positivist line on metaphysical questions. Thus the natural grouping is that of the *Aufbau* alongside the doctoral dissertation. Then, around 1929/30 there is a shift, or a branching, towards the concerns of the *Syn-*

tax. Carnap becomes interested in the philosophy of mathematics, and sustains that interest for the decade. He also becomes interested in the logical analysis of scientific theories, and the problem of concept formation. The earlier Kantian concerns now take a linguistic turn: the focus is on logic and language in science.

His first publication, a Kantstudien Ergaenzungsheft in 1922, was his doctoral dissertation *Der Raum*. By then he was 31. His masterpiece of epistemological constructionalism, *Der Logische Aufbau der Welt*, appeared in 1928. (Its first part, under the title "Konstitutionstheorie der Erkenntnisgegenstaende" had been submitted with four other pieces for his Habilitation.) By then he was 37. There followed the heyday of activities in the Vienna Circle, of which Carnap was one of the most prominent members. In 1930, aged 39, he moved to a Chair in Prague. Four years later came the publication of *Logische Syntax der Sprache*, his second Meisterwerk.

By 1936, aged 47, he had moved to Chicago. From 1939 onwards he published (with insignificant exceptions) only in English. He was 56 when *Meaning and Necessity* was published; and pressing 60 when *The Logical Foundations of Probability* appeared. All through his sixties he poured out papers on probability and induction. "The Aim of Inductive Logic" appeared just after he turned seventy.

What was the importance of his late start? And what were its effects on the character of his work?

It is easy to explain the late start. There was the Great War. Carnap served as an officer of the infantry in Berlin, apparently with light duties and having a reasonably quiet time. I say "reasonably" because he got into trouble with his high command for his political activities. He was editor of a newsletter whose editorial policy on political expression had clearly been getting out of hand. He had his knuckles rapped, and on 11 September 1918 sent out a roundrobin to his comrades saying

> Der Oberbefehlshaber in den Marken hat mir "die weitere Versendung von Rundbriefen jeder Art" verboten (Einzelkorrespondenz wird dadurch nicht betroffen.) Ich bitte deshalb, es zu vermeiden, groessere Schriftstuecke oder zusammengesetzte

> Briefe, die eine Behoerde als Rundbrief ansehen koennte, an meine Addresse zu schicken.
>
> Um so mehr nehme jeder, der in meine Naehe kommt, Gelegenheit zu Besuch und Gespraech.
>
> Ich bitte dies auch .... mitzuteilen.

Besides getting into trouble with the military authorities, he got through boxes and boxes of reading material. In 1913 and 1914 he had been a student of Frege's in Jena. This was in the middle of a period of six or seven years for which we have a detailed record of his ravenous reading. We also have evidence, from a long document entitled "Ueber Deutschlands Niederlage" that Carnap wrote in 1918 at the war's end, of his contemplative outlook on life. Grappling with the problem of his nation's honour in defeat, and with the complexities of the political situation in Europe, he turns to consider "De(n) Sinn des Geschehens." He asks

> Und welcher ist *unser* Anteil an Deutschlands Schuld? Zwar fuehlen wir uns mit dem ganzen deutschen Volke solidarisch, d.h. innerlich schicksals- und schuldverknuepft. In noch engerem Masse fuehlen wir uns aber verbunden mit denen aus dem Volke, die uns nach Lebensart, Gesinnung and Anschauung nahestehen, mit den *geistigen Menschen. Welchen Teil der Schuld tragen sie?* Ihre Gleichgueltigkeit gegen das politische Leben hat verschiedene Gruende. Von den beiden Polkraeften des geistigen Lebens: *Aktivitaet und Kontemplation,* die irgendwie eine noch unerkannte Synthese finden muessen, hat die zweite, die quietistische, mystische vielleicht auf den deutschen Menschen allzu starken Einfluss ausgeuebt.

We do not know who the "wir" of this document is: it could be a group of military comrades, an academic circle, a social class. But one is led to wonder whether Carnap's own immersion in books might not be behind his reflection that the "geistigen Menschen" had thought too much and acted too little. Carnap's record of his reading before and during the war years takes the form of two diaries: *Buecher die ich gelesen* and *Gelesene Buecher*. The detailed entries run from April 1909 to early 1917. The second diary then has a brief quantitative Zusatz taking us up to 1921, the year in which he submitted his doctoral dissertation. It was his Tante Tine's kind thought to provide the young Rudi with his first little diary on the 18th of April, 1909. After 453 entries Carnap started a newer, bigger diary in 1913, beginning with the 454th entry. The 1039th title is the last one entered. The Zusatz then reveals that sometime in 1921 he devoured item number 1417. Even for a thirty year old, this is not bad going. His reading diaries stop here, presumably because he needed more time for his other writings.

II. Early influences

What major influences do we know of, then, from this record? He had read Kant's *Prolegomena* (#247), the *Kritik der praktischen Vernunft*(#370) and the *Kritik der reinen Vernunft* (#480). Frege's *Begriffsschrift* was entry #134. The first volume of the *Grundgesetze der Arithmetik* was entry #439. The article "Funktion und Begriff" was #446. All this was before he studied under Frege, at least in the classes at Jena from which we have his lectures notes. Plato, Brentano, Fichte and Lotze make an early appearance. He read Kierkegaard in 1914; he had read a lot of Nietzsche and some Schopenhauer before that. But if the record is to be taken as complete (and what other supposition is reasonable?!), he had not, before the war broke out, read any Aristotle, Descartes, Leibniz, Spinoza, Locke, Berkeley, or Hume. His interests were *naturwissenschaftlich* and literary. He had read writers on biology such as du Bois Reymond and Haeckel. He had read Planck, Mach and Poincare. He had read Gauss and Hilbert. Lighter relief came in the form of a great deal of Goethe; several Strindbergs; Shakespeare; Schiller; Tolstoy; Rilke; Ibsen; Kipling's *Jungle Book* and Mark Twain.

III. Carnap's doctoral dissertation

The remaining major philosophical influences on his doctoral thesis he presumably encountered in his studies in the few years before: Leibniz, Husserl, Russell, Schlick and Reichenbach. His thesis is an extraordinary record of the flurry of foundational concerns in the exact sciences in the early twentieth century. Carnap had mastered the logical and mathematical works not only of figures such as Frege and Hilbert, mentioned above, but also Bolzano, Cantor, Dedekind, Einstein, Fraenkel, Hausdorff, Klein, Lobatschefskij, Lorentz, Minkowski, Pasch, Peano, Riemann, Weierstrass, Weyl and Zermelo.

The thesis itself is a masterly and all too brief summary and reconciliation of opposing philosophical views about the status of geometrical truths. It is informed by all the most recently available work in the foundations of logic and geometry. It reveals in Carnap an admirable trait which informed his personal relationships, his interest in the project of a world language, and his subsequent philosophical work: a desire for rational resolution and conciliation, and balance backed by an awareness of variation or extremes.

In the thesis one of the two original sections is a vivid example of his mathematical ability. He shows how the surface of the earth could be taken as a plane with positive overall curvature in a system of spherical geometry. Moreover, space is to be taken as homogeneous and isotropic. These conventional choices were to be offset by a further conventional choice of an unusual metric. With this, says Carnap, "muessen wir offenbar zu andern Naturgesetzen kommen, als den ueblichen" (p. 52), and he shows what these would be in the case of the principle of the conservation of energy, and Hamilton's principle. It is one of the earliest and most vivid demonstrations of the conventionalist thesis: " . . . die Bedeutung des Beispiels liegt darin, die grundsaetzliche Moeglichkeit der Wahl eines ganz andern als des ueblichen Gefueges fuer den physischen Raum zu zeigen, das aber ebenso alle Erfahrungstatsachen widerspruchslos darzustellen imstande ist" (p. 54).

He had begun his thesis with a three-fold division of space into *formal, intuitive* (perceptual) and *physical.* He concentrated on the three-dimensional cases. Within each main division he effected further subdivision: into *topological, projective* and *metric.* He deftly characterized the richer notions and structure involved with each ascent. Among the isotropic metric spaces he distinguished the homogeneous from the non-homogeneous. Among the homogeneous, characterized by their respective negative, null or positive constant curvature, were hyperbolic, parabolic and elliptical space respectively. These are the ones associated respectively with the names of Lobatschevskij, Euclid and Riemann.

He emphasized the "Wahlfreiheit" of the "Festsetzungen" concerning choice of operational metric for physical space, and choice of what is to count as straight or planar. He commends Dingler for having been the only writer to have had the insight that even the usual adoption of Euclidean geometry involves conventional choice. But Helmholz, Killing and Schlick receive credit for having stressed how changes in that conventional choice could be accommodated by changes in the formulation of laws of nature.

Carnap's discussion of formal space shows the latent logician in him. Here is where one finds the other original contribution I mentioned. He produces two quite nongeometric interpretations, or models, of Hilbert's axioms for projective geometry. One involves *strips of colours;* the other involves *consistent sets of judgments.* He emphasizes how, from the statements about these that correspond to the axioms, the two analogues of Desargues's theorem would respectively follow. This is without any apparent dependence on Bolzano's logical work, and before the rise of Tarskian model theory.

In his characterization of formal space he uses iterated class constructions. He also rehearses some of the theory of relations, order types and cardinalities. Given the infant mortality rate for foundational claims in those days, he can be forgiven the one logical slip in his dissertation. He characterizes omega as the order type of those progressions that have an initial element, and in which every element has an immediate successor and every non-initial element has an immediate predecessor. But this cannot be right; for satisfying this characterization is the order type of a copy of the naturals followed by a copy of the integers. (It seems that non-standard models of number theory dogged Carnap even unto 1940. A letter from Alonzo Church dated May 6, 1940 (008-15-17), pulls him up for a mistaken treatment of recursive definition in the *Syntax.* He draws Carnap's attention to non-standard models, and Skolem's important 1934 paper.)[1]

It is in his discussion of *Anschauungsraum* that Carnap's philosophical sympathies are revealed. He enquires after

> der logischen Begruendung der Erkenntnisse ueber den Anschauungsraum, genauer: der Grundsaetze, da die weiteren Saetze aus diesen formal-begrifflich abgeleitet werden. Erfahrung gibt nicht den Rechtsgrund fuer sie ab; die Grundsaetze sind Erfahrungsunabhaengig. . . . Denn es handelt sich hier, wie Husserl gezeigt hat, gar nicht um Tatsachen im Sinne der Ehrfahrungswirklichkeit, sondern um das Wesen (Eidos) gewisser Gegebenheiten, das in seinem besondern Sosein schon durch einmaliges Gegebensein erfasst werden kann.

His final view is that intuition gives us *a priori* the topological character of space. God, as it were, gave us the open sets; man, on the basis of his empirical experience and exercising freedom in conventional choices, did all the rest.

> Der topologische Raum stellt das ihnen allen gemeinsame dar und ist deshalb als Form des in der Wesenserschauung des Raeumlichen Fassbaren anzusehen. (p. 62). . . . Die Erkenntnis vom Gefuege des physischen Raumes ist Ehfahrungserkenntnis: sie ist gegruendet auf den >Tatbestand< der Erfahrung und wird durch Induktion gewonnen, d.h. durch Versammlung und Verarbeitung von Erfahrungstatsachen, und kann daher auch unbedingte Gewissheit nie selbst erreichen, sondern nur sich ihr als einem Grenzwert immer weiter naehern. Auf Grund des Tatbestandes ergibt sich so die Erkenntnis des topologischen Raumes, waehrend die Verwandlung dieses in eines der metrischen Gefuege nur durch Hinzunahme der frei waehlbaren Masssetzung moeglich ist. (p. 63)

What Carnap means by >Tatbestand< is "die Erfahrung, soweit sie nur in der eindeutigen, notwendigen Form vorliegt, die keinerlei frei gewaehlte Festsetzung enthaelt" (p. 65). With a finely tuned Kantian phrase, he concludes that only "die im Tatbestand enthaltenen raeumlichen Bestimmungen" can be a "Bedingung zur Moeglichkeit der Erfahrung." The axioms of formal geometry are, according to Carnap, analytic and a priori. Those of intuitive, perceptual geometry are synthetic and a priori. Those of physical geometry are synthetic and a posteriori.

The thesis has not been translated. Given its brevity and occasional rawness, it does not quite deserve to be ranked with such classics in the philosophy of mathematics as the *Grundlagen*. But it is a fascinating snapshot of the workings of a massively well-informed and fertile mind before the onset of the stirrings of logical positivism. It marks, as I have already emphasized, a late start to publication on Carnap's part. But the reading record of his diaries, the awesome *Literaturverzeichnis* and the accompanying scrupulous *Literaturhinweise* of the dissertation, all point to a bottling up of what was soon to become a majestic flow. It swept him through Vienna to Prague, on to London as the main figure of the positivist movement, then to Harvard and his first permanent position in America at Chicago. The flow was staunched only by his death.

But it was not only the studying during the war years that precipitated the belated onrush of writings. It ran in the family. The Wuppertaler Zeitung of Friday, 5th January 1934, carried on its front page an obituary for Carnap's mother on the tenth anniversary of her death:

> Zur Erinnerung an eine der groessten Frauengestalten Wuppertals. Anna Carnap geb. Doerpfeld. Geboren am 4. Maerz 1852 in Barmen-Wupperfeld, gestorben am 7. Januar 1924 in Jena. Zum 10. Todestag. . . .
>
> . . . Sicherlich ist es fuer die 35 jaehrige Anna Doerpfeld nicht einfach gesewen, aus ihrem schlichten Haushalt bei dem pensionierten Vater in Cerresheim heraus, den 61 jaehrigen Bandfabrikanten Carnap als dritte Ehefrau zu hieraten. Wieviel Taktgefuehl, wieviel Lebensklugheit, und wieviel Opfersinn gehoerte dazu! . . .
>
> Die alte Fraue schreibt eifrig an einem Manuskript, an dem Lebensbilde ihres verstorbenen Gemahls . . . Aber das Schicksal hat es anders gewollt. Nach langem qualvollem Leiden nahm ihr der Tod die Feder aus der Hand . . . Welch reiches Leben hier zu Ende ging, davon zeugen drei mit beseelter Feder geschriebene Buecher.

IV. The period before the Aufbau

It was during the time Carnap was writing his thesis that he entered into the correspondence with Dingler on conventionalism. Since Gereon Wolters has already given a comprehensive account of that exchange (in ed. N. Rescher: *The Heritage of Logical Positivism*), I shall not detail it here. But it is worth mentioning that it involved a long unpublished and untranslated piece by Carnap to Dingler on 10 August 1921, entitled "Ueber die Aufgabe der Physik, mit Anwendung auf die allgemeine Relativitaetstheorie."

All of Carnap's fascinating writings on space remain untranslated. His first published articles took up the problem of the requirement of simplicity, and the problem of dimension. Indeed, if one wishes to find a definition of the important notion of *Dimensionszahl* used in the *Aufbau,* one has to look at his paper "Dreidimensionalitaet des Raumes und Kausalitaet: Eine Untersuchung ueber den logischen Zusammenhang zweier Fiktionen." Carnap's intriguing thesis was that the fiction of the 3-dimensionality of space (hence the 4-dimensionality of the "Weltgeschehen") was the logical outcome of the latter's lawlikeness. The primary world of sensory experiences was (2+1)-dimensional. [2 for space and 1 for time]. The secondary world of physical objects and happenings, in sustaining causality, has its spatial dimension raised to 3. (The ironic inversion of Lockean terminology may even be accidental.)

These early papers on space, time and causality constitute a discernible prelude to his later epistemological concerns in the *Aufbau.* His was in effect a huge project of neo-Kantian re-construction. Kant's doctrine of space had been rehabilitated after the unsettling discovery of non-Euclidean geometries, by restricting it to the topological core common to all spaces. Time appeared to be taking care of itself, modulo Minkowski. There remained the categories. Causality revealed itself at work in the raising of dimensions. What of continuants?

It was not enough for Carnap to appeal to the spontaneous application of the category of substance. He wanted to chart a possible logical passage, via higher order abstractive definitions, from suitable primitive givens in sense experience to complex constructions that "were" the objects of our knowledge. The method was Russellian, and regrettably dated in the symbolism used. His inspiration also came from Russell. The fourth volume of *Principia Mathematica*, which never appeared, was to have been on geometry. And *Our Knowledge of the External World* was far too informal and programmatic for Carnap's tastes. Russell's "hypothetical construction" as a solution to the "problem of the connection of sense with objective reality," was, he (Russell) said,

> only in part necessary as an initial assumption, and can be obtained from more slender materials by the logical methods of which we shall have an example in the definitions of points, instants and particles; but I do not yet know to what lengths this diminution in our initial assumptions can be carried.

To this Carnap scribbled in shorthand in the margin of his own copy, "Diese Tieferlegung und Verringerung der Anfangsannahmen ist meine Aufgabe!"

The beginning was a rough typescript of July 1922 entitled "Vom Chaos zur Wirklichkeit." Later Carnap wrote on the first page "Dies ist der Keim zur Konstitutionstheorie des >Log. Aufbau<." Everything that we experience, it claimed, is only an already ordered reality. The "irrational" starting point of his theory was to be the desire to eradicate inconsistencies ("Unstimmigkeiten") in that reality by re-building it. If a well-known later figure could write that language was perfectly in order as it is, one wonders why no-one pointed out to Carnap that reality was too. But no; it was back to the drawing boards. To complete one's return to one's starting point, one had to strike out from reality every sign of order and individual determination ("Einzelbestimmbarkeit"): such as the distinction between the psychological and the physical, the ordering of the latter in space; the ordering of both these domains ("Bereiche") in time; the distinction between different sensory modalities; and the conceptual definiteness of individual sensory qualities deriving from their positions in "quality space" (for example, the position of a musical note in a scale). The result, understandably, was "Das Chaos." Having thus burned our boats, we would then have to rebuild them, while all at sea. To this end one would use "Bausteine."

These were members of the Z-series: aZb would hold just in case a and b were simultaneous (?) memories (!) and a referred (!!) to an event earlier than (!!!) b. The details are interestingly different from those later employed inthe *Aufbau.*

> Mit ihnen vollziehen wir naemlich den Aufbau. Weiter gehen wir mit der Zerlegung nicht; die Bausteine gelten uns als unteilbare Totalitaeten, obwohl sie alles das umfassen, was die spaetere Abstraktion als Gesehenes, Gehoertes usw. unterscheidet, ferner als Teiltoene eines Klanges, als die Farbflecken eines Gesichtfeldes usw. Auch wir muessen zu diesen Begriffen kommen, aber nicht durch Zerlegung, sondern durch Aufbau. Sie sind fuer uns nicht Teile von Bausteinen, sondern Klassen von solchen, die konstituiert werden durch gewisse, zwischen den Bausteinen bestehenden Beziehungen.

As this reveals, virtually the whole recipe was there as early as 1922. The paper contains much more detail which I cannot go into here, and which (as I indicated above) would bear very interesting comparison with the constructional details of the *Aufbau* of 1928, should anyone be dedicated enough to undertake it. The methodology was already well worked out in the abstract by this stage. In another published typescript, "Die Quasizerlegung" (from 1922/23), Carnap sketches the fundamentals of the method later method employed in the *Aufbau.* The subtitle is "Ein Verfahren zur Ordnung nichthomogener Mengen mit den Mitteln der Beziehungslehre."

V. Carnap's relationship with Schlick

Thus when he came to Vienna in 1925 to work under Schlick, it was with well-formed intellectual intentions. His voluminous correspondence with Schlick is a mine of material about his life and times in Vienna. It begins with polite and tentative enquiries[2] about the possibility of studying in Vienna with Schlick. In gradual stages come subtle shifts of salutation as their relationship becomes more intimate and equal. The summer of 1927 brings the "Du." Carnap was a close confidant of Schlick's. Schlick would confide about such things as an attack of colic in a lecture, or an embarassing visit by an over-exuberant and rather outrageously exhibitionist lover of his. This affair was complicated by the lover's live-in boy friend—whom I believe to have been the man who shot Schlick. [Professor Hempel and Hutten have confirmed in conversation that the rumours at the time within the Circle were to that effect.] Schlick describes[3] to Carnap how narrowly he averted a debacle when some letters were forwarded to him by Carnap from the lady in question. They arrived at the summer house where Schlick and his wife were staying; and only a "combination of luck and quick thinking" on Schlick's part prevented the disaster of uxorial interception. It was to Schlick that Carnap recounted in a letter[4] how he had married Ina (his second wife) in Prague, after a meaningless ceremony in a registrar's office conducted in Czech. This brought to a respectable end the bohemian aspect of a relationship of some years' standing; "Nun sind also meine Verhältnisse wieder in guter, bürgerlicher Ordnung."

VI. Frau X

Schlick also, from time to time, makes solicitous enquiries about one Frau *X* in Carnap's life. She was a lover and life-long friend of his. When Ina wrote to her to inform her firmly that she had taken over Carnap, but that he had her picture at his bedside and, she knew, would always love her, Frau *X* wrote back to say she was glad to learn that Ina was not jealous; for if she were, she would not be the woman for Carnap. The correspondence with Frau *X* is extraordinarily moving. It begins with a spry and youthful script on yellowed or fading pages; and it advances through the years, with the writing becoming more shaky and infirm, but clearer against newer paper. She and Carnap arranged a tryste in Berlin: a fortnight together in a *Pension,* in rooms with an interconnecting door, thoughtfully specified by Carnap in advance in order to avoid the risk of public traipsings in corridors. From that union came a daughter. Carnap wrote long and loving letters during the pregnancy. At her birth he wrote to tell his lover to teach it to speak soon: and then he would come and acknowledge it as a human being.

VII. The priority disputes with Neurath and Wittgenstein

So even by 1927 he had begun to take the linguistic turn. But this was no hard-hearted distancing of himself from his lover and their child. He was not that kind of person. The whole corpus of his correspondence bears strong witness to the contrary. The ebullient Neurath exhorts[5] "Du musst stärker sein! Carnap, Carnap werde hart?", thinking "Du musst stärker sein! Carnap, Carnap werde hart thinking Carnap's good nature too soft for the intellectual battles that Neurath saw himself as waging in alliance with him. It was Carnap who played the go-between in the battle Neurath had with Schlick over his (Schlick's) refusal to publish a submission by Neurath in the series Schlick edited with Frank. Schlick's refusal[6] had been on grounds of style and scholarship. Neurath had harangued him for over an hour in a cafe, pouring out all manner of personal reproaches and social criticism. It was Carnap who intervened by persuading[7] Neurath to make a different submission, while at the same time defending the quality[8] of Neurath's work to Schlick.

This obviously put Neurath considerably in Carnap's debt, and made it difficult to take to the patrician Schlick, the father figure of the Circle, his subsequent priority battle with Carnap. That started, in characteristic style for Neurath, with an "aufgeregtes Telegram"[9] to Frank from Moscow. It demanded acknowledgment from Carnap in the paper on protocol sentences, that he had written for publication alongside Neurath's in *Erkenntnis*. Carnap was Frank's coeditor of that isue of *Erkenntnis;* and Neurath felt[10] that he had been wronged and slighted by Carnap in that capacity, as well as in his capacities as a friend and colleague. Carnap mounted a patient, detailed and measured defence[11] of what he saw as conduct without any hint of the impropriety Neurath alleged. Neurath's frustrated shriek in his first letter[12] in reply was "Sei weniger naiv! Sei psychologischer! Sei mehr der Welt verbunden, in der Du lebst." But it is clear from the record available that Neurath never once approached Schlick about the affair; presumably because he was worried at earning a reputation for being overly combative.

Nor, for obvious reasons, does it appear that Carnap let Schlick know what had transpired with Neurath. And it was just as well. For, three months later, in August 1932, the Wittgenstein affair broke.

Why this took place will probably never become clear. Carnap offers a good deal of historical reflection on the roots of the dispute in the letters to Schlick[13] that the affair occasioned. Years later he recalled having made an entry in his diary about his first meeting with Wittgenstein: and how Wittgenstein had taken a dislike to him at the start, and had mocked him for his interest in Esperanto. But the affair of 1932 was another "Prioritaetssache." Wittgenstein had felt plagiarized, or not properly acknowledged, by Carnap in a piece on physicalism. (Wittgenstein took himself to be writing about physicalism—"nur nicht, " as he wrote to Schlick,[15] "unter diesem—scheusslichen—Namen.") His complaint to Schlick was conveyed to Carnap, who defended[16] his conduct in the same careful, measured and reasonable way that he had employed with Neurath. This was relayed back to Wittgenstein, who then wrote[17] to Carnap a fey, shrill and petulant letter, which he had Schlick forward on his behalf. In this letter he ends by saying

> Mein Zweck ist es nicht, die Nennung meines Namens in Ihren Schriften zu erreichen . . . ; ja ich hoffe, dass ich bald auch nicht mehr in der—wenn auch lockeren—Beziehung zum akademischen Leben stehen werde, wie zur Zeit. Aber ich empfaende es als Unrecht, Sie ueber meine Ansicht in der Sache nicht voellig zu unterrichten. Hochachtungsvoll, L.W.

Wittgenstein's complaint was that he, as allegedly the "Hauptquelle" of the ideas expressed in Carnap's piece, had not been named; although Neurath had been cited, as well as earlier publications by Carnap. (Carnap underlined the occurrence of "Hauptquelle" in his copy [18] of Wittgenstein's letter to Schlick, and placed an exclamation mark beside it.) Wittgenstein accused Carnap of having harboured the deliberate intention not to cite him, and to conceal "die Provenienz der Gedanken." The view that Schlick took[19] was that Carnap ought to have made some appropriate acknowledgement. For Schlick's embarrassed position was that Wittgenstein was a frequent summer visitor to his Circle, communicating ideas that were in long gestation before appearing in print. Wittgenstein's oral publication had lost him his claim to priority. And indeed in subsequent correspondence about the *Syntax* Schlick was moved of his own accord to insist[20] to Carnap that proper acknowledgement be made of Wittgenstein's views. But Schlick, in the original brush between the two men, also defended Carnap to Wittgenstein, maintaining that Carnap had been "reines Herzens," acting only negligently rather than fraudulently.

Carnap himself later reflected, in a piece not intended for publication or even circulation (RC 102-78-08, dated 16.11.56, entitled "Dies *nicht* im Autobiographie!"):

> Ich glaube, ich war der erste, der Wittgenstein im deutschen Sprachgebiet bekannt machte. Als ich aber einmal (in "Die physikalische Sprache als universale Sprache der Wissenschaft") ihn nicht nannte, weil der physikalistische Standpunkt, den ich dort vertrat, stark von dem von W. abwich, schrieb er einen sehr indigneirten Brief (30.8.32 in Prag bekommen) an mich, den er an Schlick sandte mit der Bitte, ihn mir weiter zu schicken. Bis dahin hatte ich einen hohen Respekt auch vor W.s menschlichem Charakter gehabt; aber dieser Brief war so voll von Hass, ausgedrückt in einem unglaublichen, persönlich gehässigen Ton dass ich tief schockiert war . . . Jahre später baten einige der *Studenten Wittgensteins in Cambridge* um seine Erlaubnis, Transkripte seiner Vorlesungen an Freunde und interessierte Philosophen zu schicken. Er fragte nach der Liste der Namen, und dann *billigte er alle ausser meinem.* In meinem Leben habe ich niemals auch nur etwas ähnliches erlebt wie diesen Hass gegen mich. Ich habe keine hinreichende Erklärung; die hätte wohl nur ein Psychoanalytiker geben können. (Carnap's emphases)

VIII. After Vienna

It is impossible to say what Schlick's view might have been, and what might have been the consequences for Carnap's career and a good deal of twentieth century philosophy subsequently, had the Neurath complaint reached him in time to colour his opinion when it came to Wittgenstein's complaint. Certainly Carnap owed a great deal to Schlick's intellectual patronage in the then highly structured and testimonial-dependent[21] world of academic publishing and university promotions. But by this time Carnap was in Prague, having left as an acknowledged leader of the Positivist movement. The *Aufbau* had been published, and work on the *Syntax* was well under way. He had been editing *Erkenntnis* with Reichenbach. Various letters[22] to him testify to how sorely he was missed when he left. Apparently the meetings became less lively and more stilted, under Schlick's increasingly formalized chairmanship. It was this period of Circle meetings which Ayer attended, and which he describes in his autobiography.

Although Carnap did attend many meetings of the Circle after his move to Prague, he appears to have missed the very first Circle discussion, with Goedel present, of his incompleteness theorems. This was in January 1931, shortly after Carnap had left for Prague. But Carnap had already had it from the horse's mouth. His *Aufzeichnungen* contains the following entries:

> VIII/1930
> *Di 26* . . . 6½ 9 Cafe Reichsrat mit Feigl, Goedel, dann Waismann. Reiseplan mit Schiff nach Koenigsberg. Goedels Entdeckung: Unvollstaendigkeit des Systems der PM; Schwierigkeit des Widerspruchsfreiheitsbeweises. . . .
>
> Fr 29 . . . 5-9 Cafe Reichsrat. Zuerst erzaehlt mir *Goedel* von seinen Entdeckungen. Dann frage ich ihn wegen Verzicht auf verzweigten Typentheorie . . .

Minutes of meetings held shortly after Goedel's presentation show Carnap taking an active part, with his own work as a major focus for discussion by others. In June 1931 he gave three papers in succession on metalogic. Earlier that year they had been discussing his thesis of physicalism, and had heard his "Geaenderte Auffassung gegenueber dem logischen Aufbau der Welt."

But even Carnap could not re-build the world around him. The storm clouds were gathering, and he was already looking, with the help of Nagel and Quine, for an opening in the United States. He had been there once before. A letter to Dingler dated 9 May 1923, written on a train from New York to Mexico, records this impression:

In den Ver. St. sind wir nicht lange gewesen. So haben wir zwar nur einen fluechtigen, aber doch lehrreichen Einblick getan. Wir Deutschen koennen uns kaum mehr vorstellen, was ein nicht durch den Krieg verarmtes Volk alles schaffen kann."

Nagel came to visit him in Prague. There is a delightful exchange (029-05) between Nagel and the Carnaps, much of it written by Ina, and full of her dry witticisms.[23] Nagel writes[24] to confess that he had to miss a rather interesting meeting of the Circle. He was off to Genoa to make a particularly attractive rendezvous. The meeting he missed was one where Tarski and Goedel were discussing Carnap's paper on the antinomies. "That," wrote Nagel, "[is] what comes of having a private life."

Later, once back in America, Nagel wrote to suggest that much could be done to combine elements of American pragmatism with logical positivism. Carnap also reported in a letter to Schlick (029-28-16, 13 May 1934) that Lewis had written to him "Some day I hope that some member of your group may publish a paper in which some comparison may be made between your views and the pragmatism of Peirce and James and Dewey."

Another visitor to Prague was Quine. Through his assiduous help, to which his generously contributed complete record of his correspondence with Carnap bears ample testimony, Carnap was encouraged to make the move to America.

The rest, as we all know, is more recent intellectual history.

POSTSCRIPT

I am grateful for having had the privilege of being able to talk on this topic to the colleagues gathered to celebrate the Center's 25th anniversary. It came close to carrying coals to Newcastle. Some present had been Carnap's close associates and friends. Some had edited his works posthumously. Yet others had ensured that the rich store of first editions, correspondence, unpublished notes and manuscripts, mementoes, marginalia and other memorabilia have now found a superb home in the Hillman collection. I was fortunate to gain access to this collection in March and April 1985, as a visitor to the Center. To those involved, I owe thanks for extracting gratitude in this way.

I owe special thanks to Richard Nollan, the archivist and curator of the collection. Much of what I have to say is merely colourful collage, impressionistic re-arrangement, of material whose historical and philosophical significance he had noted beforehand. The philosophical community will be in his debt for his painstaking transcription of Carnap's screeds of shorthand.

This time is ripe for *two* books on Carnap: one on the man and his life; the other on his work. (Perhaps also a third: one making available in English translation for the first time all the important early papers on which his considerable reputation was built.) But for those interested in writing on the man and his life, there are constraints of confidentiality on one's use of the collection. These are not simply matters of legally enforceable print. After immersion in the intimacies of the *Nachlass,* they become constraints of conscience. As soon as one enjoys the privilege of sharing one feels one has trespassed. Yet there are others who hold that, in the highest interests of intellectual inquiry, it should all be in the public domain. Perhaps, as the parties most concerned by disclosures pass away, this view will prevail. In the meantime we are on middle ground; and I have tried to strike a balance. I have reported only in the most general terms such intimate information as may have shed light on my subject.

NOTES

1. "Uber die Nicht-charakterisierbarkeit der Zahlenreihemittels endlich oder abzählbar unendlich vieler Aussagen mit ausschliesslich Zahlenvariablen" (*Fundamentae Mathematicae*).

2. 029-32-50, 9 August 1924 (from Schlick); 029-32-48, 11 November 1924 (to Schlick).

3. 029-32-03, 5 August 1927.

4. 029-28-31, 5 March 1933.

5. 029-14-01, 20 December 1930.

6. Letter to Frank dated 16 July 1930 (029-30-07) with report (029-30-08).

7. Letter dated 25 July 1930 (029-14-11).

8. Letter dated 28 August 1930 (029-30-04).

9. See Frank's postcard to Carnap, dated 3 February 1932 (029-12-67). The telegram is 029-12-68.

10. 029-12-59, 10 March 1932.

11. 029-12-63, 10 February 1932; 029-12-60, 2 March 1932.

12. 029-12-62, 17 February 1932.

13. See especially 029-29-04, 28 September 1932.

14. RC 102-780-07 Über Wittgenstein, aus meinem Tagebuch 1927.20.6.

15. 102-78-02, 8 August 1932, p. 2.

16. 029-29-09, 17 July 1932, 029-29-04, 28 September 1932.

17. 102-78-03, 20 August 1932.

18. 102-78-02, 8 August 1932, p. 1.

19. This view he had first expressed in a letter dated 10 July 1932 (029-29-10), more than two months before Wittgenstein's letter(s) reached Carnap.

20. Letter dated 2 June 1934 (029-28-13).

21. See e.g. 029-32-08, 31 March 1927 (Schlick's testimonial regarding the *Aufbau*).

22. See e.g. Nagel's letter dated 5 January 1935 (029-05-16).

23. See especially 029-05-15, 23 January 1935.

24. 029-05-14, 6 March 1935.

MOWING THE GRASS THAT'S GRUE: WHAT CARNAP SHOULD HAVE SAID TO GOODMAN

RICHARD CREATH

We gather today to celebrate an extraordinary event, the founding of the Center for Philosophy of Science twenty-five years ago, and also, of course to celebrate the continuation of that extraordinary institution as well. I, however, am here to worry about something very ordinary indeed. Each spring, summer, and fall day, for the past twenty-five years that the Center's first Director, Adolf Grünbaum has looked out his office window toward Schenley Park, he has seen that the grass is green. Now it may be an extraordinary event to be able to see at all through the grime that would appear to be an essential feature of windows in the Cathedral of Learning, but it is completely ordinary for the grass to be green. What is also perfectly ordinary is for Professor Grünbaum to infer that he will likewise find the grass in Schenley Park to be green on the next spring, summer, or fall day that he observes it. None of this would be worth remarking on were it not for certain very well-known difficulties that were discovered by Nelson Goodman and that have become known as the new riddle of induction or more popularly as the paradox of "grue" and "green."

Robert Nozick once described Marxism as normative sociology,[1] i.e., the sort of study which attempts to find out what the causes of social problems *ought to be*. Well, the remarks that follow might be described as normative history of philosophy; my subtitle after all is "What Carnap Should Have Said to Goodman." Okay, normative it may be, but it is nonetheless highly constrained by historical fact. The enterprise I have in mind involves, first, discovering what was said and why that is problematic. Second, it includes developing some alternative that would be better. In both parts of this we must try to see the issues from Carnap's point of view. That is what makes normative history so hard: in order to carry out the corrective part of the enterprise I will have to show that the changes I am proposing are consistent with Carnap's most fundamental commitments. And finding out where those commitments lie requires careful sifting of evidence as well as judicious assessments of what depends on what. In short, the task is plainly historical, in the best sense, and if it succeeds it will reveal the core of Carnap's philosophy as no mere chronicle could.

The paper that follows comes in two parts along the lines just sketched. Its first half traces the exchange between Goodman and Carnap on the new riddle of induction. In order to see what prompted all the fuss, our examination begins with Goodman's original query. I shall then look at Carnap's reply and at Goodman's rejoiners, both immediately and in *Fact, Fiction and Forcast.* In each case we will try to see both what the content is of the proposed solution and why that solution will not work. In the last half of the paper I present an approach to the problem that I think Carnap *should have* adopted. I shall argue that this approach would have allowed Carnap to deal effectively with the new riddle of induction within the framework of Carnap's own epistemology. Further, I go on to argue that the proposed approach would have the additional virtue of allowing Carnap to identify and absorb what is right about Goodman's entrenchment solution. These provide, I think, good reasons to believe, not perhaps that we must accept the proposed solution, but at least that Carnap should have.

Goodman's first paper on the topic was "A Query on Confirmation,"[2] and it dates from 1946. There he raised several problems which are far as he could tell infected every available theory of confirmation including those of Oppenheim, Hempel, Carnap and Helmer. Most of the problems Goodman discussed there have quietly been put aside, but one has endured. Even here the precise example has been altered over the years, though not its structure nor the threat poses to simple induction. Here I'll use the later version so as to conform to current practice. Briefly the difficulty is this: if all emeralds examined before a certain time *t* are found to be green, then at *t* our observations would seem to support the claim that the next emerald seen will be green. But consider the predicate "grue" which

> applies to all things examined before *t* just in case they are green but to other things just in case they are blue. Then at *t* we have, for each evidence statement asserting that a given emerald is green, a parallel evidence statement asserting that that emerald is grue.[3]

This evidence should support to an equal degree the claim that the next emerald seen will be grue. Obviously these two supported claims are incompatible. Unfortunately the difficulty is readily generalizable, that is for any claim inductively supported by any evidence there would seem to be another claim incompatible with the first but which is inductively supported by equal and parallel evidence.

Goodman's paper was explicitly a query or a request for clarification rather than a refutation, and it was by no means directed specifically at Carnap. Nevertheless, Carnap feared that Goodman's paper would be seen as a criticism primarily of Carnap's own emerging theory of probability. In consequence, in the correspondence between himself and Goodman, though Carnap initially welcomed Goodman's discovery, he went on quickly to become uncharacteristically nettlesome and defensive on the matter.[4] In fact he became so vexed that he published a reply, "On the Application of Inductive Logic.[5] This, of course, had the unintended but inevitable result of reinforcing any suggestion that Goodman's paradox threatened Carnap's views in particular.

Briefly, Carnap's reply is that the problems Goodman raises are problems of application only and not difficulties within the theory. The difficulties arise only because inductive logic is at an early stage of development and has been worked out only for simple languages. Even so these languages are richer than those to which deductive logic was confined for over two thousand years. Moreover, beginning with simple languages is methodologically correct, just as it is correct for physicists to begin to apply their theories in artificially simple situations (such as in experiments involving billiard balls). The physicists, too, must make dubious (even false) but methodologically useful simplifying assumptions.

The simplifying assumption that Carnap makes, however, is powerful indeed, for he presumes that there are logically independent, logically simple predicates into which everything that is sayable in the language may be analyzed. This underlies his notion of state description, which in turn is at the heart of his inductive logic. Now with a notion of state description in hand Carnap could (and did) recast his deductive logic in terms thereof. He even came to believe that this was the only way of doing things and that the assumption of logically simple predicates was essential for deductive logic especially for determining the logical independence of predicates. Of course Carnap was mistaken, and this mistake is some indication of how deeply entrenched this state description view had become in Carnap's thought.

If there are logically simple primitive predicates than all predicates can be classified into one of three categories:

(1) *Purely qualitative:* those that can be expressed without individual constants.
(2) *Purely positional:* can be expressed without using any primitive predicates [where '=' is taken by Carnap as a logical sign rather than as a two-place predicate.]
(3) *Mixed predicates;* neither (1) nor (2)

How does this tripartite division help us discover which predicates are inductively projectible? First, we have to decide what we shall mean by "inductively projectible." Carnap offers the following definition (phrased in material mode): a property W is *inductively projectible* if it is always the case that the higher the observed frequency of W in an observed sample, the higher is, on this evidence, the probability that a non-observed individual has the property W. Carnap then makes the conjecture that all and only purely qualitative predicates are inductively projectible.

Whether it is necessary to mention such a restriction explicitly is another matter. According to Carnap those, such as Laplace and

Reichenbach, who have formultated inductive rules without specifying to which properties those rules were to be applied probably intended them to apply only to purely qualitative properties. Still, a complete presentation of such inductive rules would require explicit mention of the restriction.

For himself, on the other hand, Carnap believed that no further rule was required to specify the projectible properties. This is because he had given an explicit definition for the degree of confirmation C^*. Given the definition for C^* and given the definition for inductive projectibility, it is a purely technical matter whether a certain property is inductively projectible in his system. Purely technical it may be, but it might also be a hideously complicated technical exercise to prove that a given property is or is not inductively projectible. Presumably, it was for this reason that Carnap left unproved his conjecture that in his system all and only purely qualitative properties will turn out to be inductively projectible.

Summarizing, then, Carnap proposes to meet Goodman's challenge in this way: All predicates must be resolved into a set of logically simple, logically independent primitive terms. This allows Carnap to distinguish between "green" and "grue," i.e., he can say that the former is purely qualitative while the latter is not. Carnap's insistence on an analogy between inductive and deductive logic is an attempt to make the strategy of resolving things into simple predicates look benign, even unavoidable. Finally, Carnap's claim that the question of which predicates are inductively projectible is a purely technical matter is part of an attempt, not to distinguish the projectible from the non-projectible predicates, but to defend himself from the charge that he had omitted such a distinction from previous presentations of his theory.

I don't like Carnap's answer precisely because it assumes that some predicates are *really* simple without in any way relativizing this to a language. I think it's downright un-Carnapian. I suspect that Carnap thought so, too, for in a later note of the exchange with Goodman he said:

> I must confess that I, too, have a rather uneasy feeling concerning the concepts of absolute simplicity and absolute completeness referred to in the requirements. I hope very much that it

> will be possible to find a way of avoiding these problematic concepts and replacing them by the kind of relative concepts with which we usually work. But at the present moment I do not see whether or how this can be done.[6]

Even if Carnap was uneasy with it, his attempt to distinguish "grue" from "green" on the grounds that the latter but not the former is purely qualitative has enormous intuitive appeal. Carnap was by no means the last to try to answer along these lines.[7] After all, besides the obvious fact that "grue" refers to or intimately involves an individual, namely *t*, what intrinsic difference could there be between the two predicates?

Goodman eventually provided his own answer, of course, but first he had to demolish Carnap's answers. Initially, Goodman only protested against any attempt to make simplicity absolute.[8] But in *Fact, Fiction and Forecast* Goodman was able to show that there is an exact symmetry between "grue" and "green" that menaces any attempt to distinguish them. While it is true that "grue" may be defined in terms of "green," "blue," and (a name of) an individual instant *t*, it is equally true that "green" and "blue" may be defined in terms of "grue," "bleen," and (a name of) the individual instant *t*. If the definability of "grue" shows that it is not purely qualitative, then the corresponding definability of "green" must show that it is not purely qualitative either.

In the wake of *Fact, Fiction and Forecast* there arose a cottage industry[9] of attempts, ranging from the inspired to the idiotic, to distinguish "grue" and "green." If he had done nothing else, Goodman could have made a handsome career just from writing replies. He did write replies, too, until he tired of it, but that need not concern us here.

What does concern us is Goodman's own solution, and that turns on the notion of entrenchment. As Goodman tells us:

> Plainly "green," as a veteran of earlier and many more projections than "grue," has the more impressive biography. The predicate "green," we may say, is much better *entrenched* than the predicate "grue."[10]

Plainly, this is at least some difference between the predicates, but there are those who doubt that a difference of entrenchment is sufficient or even the right sort of thing to warrant projecting "green" but not "grue." Surely, the criticism goes, a predicate becomes entrenched because it is the right kind of predicate rather than it being the right kind because it becomes entrenched. I'm willing to postpone judgment temporarily on this criticism because I think Goodman's proposal is far more seriously defective. As I have shown in more detail elsewhere[11] it is doubtful that in crucial cases a difference in entrenchment can even be detected. If not, Goodman's inductive theory would be inapplicable; unless we can detect which predicates are entrenched, we cannot use entrenchment theory to decide which predicates to project. Note that if my objection is correct, it poses the same threat to Goodman's inductive theory that Goodman's original puzzle posed to antecedent inductive theory: if the objection stands the inductive theory in question (Goodman's in this case) cannot be used.

So why can't the greater entrenchment of "green" over "grue," supposing there be such, be detected? Certainly the word 'green' has been used in projections more often than the word "grue." But Goodman carefully and correctly distinguishes the word from what gets entrenched, namely a class. As Goodman himself puts it:

> In a sense, not the word itself but the class it selects is what becomes entrenched, and to speak of the entrenchment of a predicte is to speak elliptically of the extension of that predicate. Moreover, no entrenchment accrues from the repeated projection of a word except where the word has the same extension each time.[12]

Remember, now, that two classes in question differ at t only with respect to their *unexamined* members. A pre-t projection of one class would be behaviorally indistinguishable from a pre-t projection of the other class. We cannot guarantee that a past projection of the word "green" is a projection of the class selected by my word "green" rather than that selected by my word "grue." Worse, not only can we not guarantee at t a claim, at t we cannot even confirm it without confirming the corresponding alternative precisely as much. Of course, after *t* we might be able to tell which class had been projected, but this would be no help at *t*. Even this would require somewhat doubtful assumptions about constancy of reference. Moreover, since "grue"-like predicates can also be constructed for later times we would be no nearer a solution for the general problem that Goodman poses.

Finally even if we suppose that some predicates are more "natural" than others, in the sense that they are more likely to be acquired than others, this would not help Goodman's cause. If we can in fact discover how natural a given predicate is, then we can resolve in cases of conflict to project the more natural predicate and forget all about entrenchment.

So, all in all, things look as bad for entrenchment theory as they did for Carnap's attempt to find purely qualitative predicates. On the other hand, without entrenchment what shall we do about the new riddle of induction? In what follows I want to advance a solution more Carnapian than Carnap's that I think he should have adopted. This is the normative part of our normative history of philosophy. To see how this might be so let us return to our original case.

Figuring 150 spring, summer, and fall days per year that Professor Grünbaum enters his office and looks toward Schenley Park, in twenty-five years that makes 3750 times that he has done this. Each time he has seen the grass to be green. But since the literature on this topic goes back to the 1940's and Professor Grünbaum is well read, let us suppose that each time he was also fully aware that the grass was grue (where the '*t*' in the usual definition thereof is, say, December 13, 1985). If Professor Grünbaum has powerful inductive evidence that the next non-winter Schenley Park grass he sees will be green, he would seem to have equally powerful evidence that that grass will be grue. If both "green" and "grue" are available for projection we get incompatible results. This is the first problem generated by Goodman's paradox.

The problem of incompatible projections has suggested to some, Goodman apparently included, that we must find some absolute difference between "grue" and "green." Considered merely in their intrinsic character as classes there would appear to be no such difference, so he was forced to look to their respective historical associations. Contrary to this view, however, avoiding the problem of incompatible projections does not require that there by any absolute difference between the predicates, but the simultaneous uses of both predicates must indeed be prevented.

In his symetry argument in *Fact, Fiction and Forecast* Goodman in effect proved that neither "grue" nor "green" was purely qualitative with respect to the other. This suggests that we might relativize projectibility to a language system: we might insist that within any system all predicates must be purely qualitative with respect to each other. In the present case this would yield two different language systems, the first containing "green," "blue," and the other familiar color words and the second containing "grue," "bleen," and the rest of the Goodmanian spectrum. Actually, of course, there would be indefinitely many such systems, but we may confine our attention to these two. The important point is that as long as we stay solely within any one of these systems the problem of incompatible projections does not arise.

I am not here attempting the technical problem of specifying what it is for a family of predicates to be purely qualitative with respect to each other. Carnap's definition of a purely qualitative predicate might serve as the basis for such a definition within a restricted family of languages, but that definition would need to be generalized. Goodman, in any case, has never challenged the idea that a definition of qualitativeness with respect to a system can be provided. His argument was against qualitativeness *simpliciter*, and the argument was to the effect that the notion must be relativized to a system.

There are two objections to this proposal that should be addressed. The first is that it would disallow such predicates as, for example, "arctic" and "solar" that we currently use and project harmlessly even though they, intuitively, refer to particular individuals or locations.[13] Here it seems that other purely qualitative predicates can be found to do all of the inductive work of these predicates, and that more generally non-qualitative one-place predicates can be supplanted by qualitative multi-place predicates. The use of predicates such as "artic" and "solar" arises no doubt because the phenomena to which they apply are fairly restricted. There is no *a priori* reason why this need be the case, and hence no reason why the predicates used should not be qualitative.

The second objection is that the proposal avoids incompatible projections at the cost of arbitrariness about which language system to use. This problem is so large and important and goes so directly to the heart of both Carnap's and Goodman's discussions that we must examine it carefully. Indeed, we may consider the problem of arbitrariness as the second major problem to be raised by Goodman's new riddle of induction. Here, it would seem, that we may need to distinguish absolutely between "grue" and "green." This, of course, is what Goodman's theory of entrenchment and Carnap's theory of pure qualitativeness were supposed to do. While for the reasons given earlier, I do not believe that either theory is satisfactory, I think there is a much deeper current in Carnap's thought that will allow us to deal with the problem of arbitrariness.

We can get at this deeper current by looking at the moral that Goodman himself draws from his own paradox. The paradox, he says,[14] shows that there are two features of a valid induction: first, there must be the right formal relation between premises and conclusion, and, second, there must also be the right categorization in the formulation of the premises. Goodman is right, of course, and Carnap could readily have agreed. But Goodman's story even on the rightness of relation between premises and conclusion differs importantly from Carnap's.

According to Goodman, the right formal relations are those embodied in an *explication* of ordinary inductive (and deductive) practice. There is *one* unique set of correct rules or perhaps at most a small group of such sets. Ordinary practice is invested with special authority. Our ordinary beliefs, including beliefs about justification, have what Goodman call "initial credibility."[15]

For Carnap, by contrast, the choice among alternative sets of rules (inductive and deductive) is *conventional,* though guided by *pragmatic* considerations. There is no set of rules that is uniquely correct, epistemically speaking. Explication tells us what ordinary practice is, but it has no special authority. Faced with a choice between an explication of ordinary practice and an alternative that is pragmatically more useful, the practical decision, according to Carnap, should be to supplant ordinary practice.

This conventionalist aspect of Carnap's thought is important and too little studied. As part of a book in progress on the Carnap-Quine debate over analyticity I document Carnap's conventionalism, but my point today is that if Carnap embraced conventionality with respect to the inductive rules, he should likewise have done so with respect to the inductive categories. In other words the suggestion is that the second problem generated by Goodman's paradox, the problem of arbitrariness, is not so devastating. We *can* learn to live with it. Consider, for example, that alternative sets of inductive rules yield incompatible projections, too. This is not disasterous if we don't use rules from alternative sets simultaneously. Moreover, one cannot guarantee in advance that inductive rules will give correct predictions; why should more be demanded in advance of the inductive categories. The choice of such categories, like that of inductive rules, should be guided by pragmatic utility. Of course, among certain specially contrived alternatives there is as yet no difference of track record, so we have as yet no reason to believe one more useful than another. Suppose it could be shown that the primitive predicates of one system could readily be learned as observation predicates (i.e., that we could easily learn to use it in reporting our observations) while the primitive predicates of another system could not be so learned. This might then serve as a pragmatic reason for preferring one system to the other. This would be one way of viewing Wesley Salmon's argument in "On Vindicating Induction" that "green" but not "grue" is ostensively definable.[16] Even on the conventionalist view I am advancing we should not be surprised to discover strong, even compelling, pragmatic arguments in favor of some systems over others. Whether Salmon's argument will ultimately succeed, however, I cannot say. We are, after all, creatures with our own internal biological clocks, and I am not yet prepared to say what is possible in the way of ostensive definition. Whatever the outcome, Salmon does seem to be offering the right sort of argument from Carnap's point of view.

What then am I offering on Carnap's behalf (even if against his example)? Essentially the proposal is to divide Goodman's new riddle into two parts: a problem of incompatible projections and a problem of arbitrariness. The first is resolved by relativizing projectibility to a system, and the second is treated by arguing that we can learn to live with it; we have to anyway. Thus, Carnap could have responded effectively to Goodman's riddle while remaining consistent with Carnap's own epistemic views.

This, however, is not the end of the matter. The proposed solution has another virtue, namely it would allow Carnap to capture what is right about Goodman's entrenchment solution. We noted earlier that one strong body of opinion is that even if entrenchment can be detected, it is a symptom rather than a cause of the rightness of categorization. Goodman generally refuses to defend himself on this point, and, perhaps an attitude of "any port in a storm" is not unreasonable. Though the proposal here suggests that there may indeed be other "ports" than entrenchment theory, entrenchment itself need not be rejected.

But for the Carnapian it would have a different role than it does in Goodman's theory. For him or her there can be no *epistemic* explanation of the absolute rightness of "green" over "grue." But there can be all kinds of other differences between the predicates as indeed there are. One such difference is our overwhelming preference for "green" over "grue." If the difference between them is conceived in this way entrenchment could be part of a psychological or sociological explanation of how we come to have this overwhelming preference. In view of the difficulties lately noted about detecting entrenchment, however, we may find that entrenchment is a property of words as much as it is of classes.

While for the Carnapian entrenchment is a psychological rather than an epistemic notion, it should be noted that it is unclear whether Goodman even *wants* to separate epistemic from psychological or sociological sources of our epistemic views. Whether he does or not, in some sense he has the order right: at least the feeling of rightness of categorization does derive from, rather than underlie, entrenchment.

In summary, I am suggesting that Carnap's own answer to Goodman's "Query on Confirmation" was not sufficiently Carnapian. By appealing to the absolute simplicity of certain predicates he neither satisfied Goodman nor remained consistent with his own deepest beliefs. Instead, he should have responded to the problem by dividing it into two problems. He should have dealt with the problem of incompatible projections by relativizing projectibility to a system each of whose basic predicates is purely qualitative with respect to all the others. He should have dealt with the problem of arbitrariness by showing that the choice of inductive categories is no more arbitrary than the choice of inductive (or for that matter deductive) rules. Carnap was prepared, not only to tolerate, but to embrace that level of arbitrariness. This approach develops what I take to be the deepest themes in Carnap's epistemology, and absorbs (after a fashion) what Carnap did say in response to Goodman. The suggestion also has the virtue of doing justice to the widespread conviction that entrenchment cannot be a source of epistemic virtue as well as to Goodman's own view that entrenchment *preceeds* our epistemic conviction that "green" is a right category for projection.

There seems to me no doubt that this is what Carnap should have said. Our normative history of philosophy carries us that far. But it does not follow that we must agree with Carnap. To the extent that we work out our own alternative epistemic views that match in depth and scope those of Carnap, we shall devise alternative answers of our own to Goodman's riddle. In the meantime it is reassuring to know that Carnap could have answered Goodman, well within the framework of his own epistemology.

But what then of Professor Grünbaum? When last we left our story he was pondering what color the grass would have on the 3751st non-winter day that he looks out of his office. On the Carnapian story here offered, Professor Grünbaum gets to choose his language system and the correct projection is settled thereafter. After twenty-five years I think he deserves the luxury of choice. Since his preferred language system is ordinary English which he employs with both power and style, I need not tell you what projection he'll make.[17]

NOTES

1. Robert Nozick, *Anarchy, State, and Utopia* (New York: Basic Books, 1974), p. 247.

2. Nelson Goodman, "A Query on Confirmation," *Journal of Philosophy*, vol. 43 (1946), pp. 383–385.

3. Goodman, *Fact, Fiction, and Forecast* (Indianapolis, IN.: The Bobbs-Merrill Co., 1965), p. 74.

4. Carnap also remained personally friendly with Goodman and supportive of Goodman's work. See especially manuscripts [084-79-02], [084-19-06] pp. 1–3, and [084-19-26] in the Rudolf Carnap Collection, Archive for Scientific Philosophy, Special Collections Department, Hillman Library, University of Pittsburgh. I would like to express my appreciation to Richard Nollan, Charles Aston and other librarians at the Archive for their help.

5. Rudolf Carnap, "On the Application of Inductive Logic," *Philosophy and Phenomenological Research,* vol. 8 (1947), pp. 133–48.

6. Carnap, "Reply to Nelson Goodman," *Philosophy and Phenomenological Research,* vol. 8 (1948), pp. 461–2.

7. See, for example, Stephen Barker and Peter Achenstein, "On the New Riddle of Induction," *The Philosophical Review,* vol. 69 (1960), pp. 511–22. In fact, Carnap himself repeated what is essentially the answer of "On the Application of Inductive Logic" in "A Basic System of Inductive Logic, Part I" (especially pp. 73–4) which appears in Rudolf Carnap and Richard C. Jeffrey (eds.), *Studies in Inductive Logic and Probability,* Vol. I, (Berkeley, CA: University of California Press, 1971), pp. 33–165.

8. Goodman, "On Infirmities of Confirmation Theory," *Philosophy and Phenomenological Research,* vol. 8 (1947), pp. 149–51.

9. See Goodman, "Reply to Comments on *Fact, Fiction, and Forecast,*" in *Problems and Projects* (Indianapolis, IN.: The Bobbs-Merrill Co., 1972), pp. 398–412 for Goodman's remarks on some of the more significant of this literature.

10. Goodman, *Fact, Fiction, and Forecast, op cit.*

11. Richard Creath, "A Query on Entrenchment," *Philosophy of Science,* vol. 45 (1978), pp. 474–77.

12. Goodman, *Fact, Fiction, and Forecast, op cit.*, pp. 95–6

13. Goodman, "On Infirmities of Confirmation Theory," *op cit.*, pp. 149-50.

14. Goodman, "Notes on the Well-Made World," *Erkenntnis,* vol. 19 (1983), pp. 105–6

15. Goodman, "Sense and Certainty," *Philosophical Review,* vol. 61 (1952), pp. 160–67.

16. Wesley Salmon, "On Vindicating Induction," in Henry E. Kyburg, Jr. and Ernest Nagel (eds.), *Induction: Some Current Issues* (Middletown, CT.: Wesleyan University Press, 1963), pp. 27–41.

17. I would like to thank Professors Israel Scheffler and Catherine Z. Elgin for comments on earlier formulations of these ideas.

CONTRIBUTORS

SYLVAIN BROMBERGER
Department of Linguistics and Philosophy
Massachusetts Institute of Technology

MARIO BUNGE
Foundations & Philosophy of Science Unit
McGill University

RICHARD CREATH
Department of Philosophy
Arizona State University

ARON EDIDIN
Department of Philosophy
University of Notre Dame

BAS C. van FRAASEN
Department of Philosophy
Princeton University

J. A. FODOR
Department of Linguistics and Philosophy
Massachusetts Institute of Technology

RONALD GIERE
Department of History and Philosophy of Science
Indiana University

ERNAN McMULLIN
Department of Philosophy
University of Notre Dame

MICHAEL L. G. REDHEAD
History and Philosophy of Science
Chelsea College

ALVIN E. ROTH
Department of Economics
University of Pittsburgh

HERBERT SIMON
Computer Science/Psychology Dept.
Carnegie Mellon University

BRIAN SKYRMS
Department of Philosophy
University of California

NEIL TENNANT
Department of Philosophy
Australian National University

NAME INDEX